GNU FisicaLab Reference Manual

A catalogue record for this book is available from the Hong Kong Public Libraries.

Published in Hong Kong by Samurai Media Limited.

Email: info@samuraimedia.org

ISBN 978-988-8381-78-4

Background Cover Image by https://www.flickr.com/people/webtreatsetc/

Table of Contents

1 Introduction **1**

   1.1   Handling the elements 3

   1.2   Element data 4

   1.3   How it works 7

   1.4   Messages 7

2 Module kinematics of particles **8**

   2.1   Stationary reference system 10

   2.2   Mobile reference system 11

   2.3   Mobile reference system in X/Y 12

   2.4   Mobile 13

   2.5   Mobile in X/Y 14

   2.6   Mobile in X/Y with constant velocity 15

   2.7   Cannon 16

   2.8   Mobile radial 17

   2.9   Distance 18

   2.10   Distance XY 18

   2.11   Point 19

   2.12   Relative velocity 19

3 Examples kinematics of particles **20**

   3.1   Example 1 20

   3.2   Example 2 22

   3.3   Example 3 24

   3.4   Example 4 26

   3.5   Example 5 29

   3.6   Example 6 31

   3.7   Example 7 33

4 Module circular kinematics of particles **35**

   4.1   Stationary reference system 37

   4.2   Mobile with circular motion 38

   4.3   Mobile with polar circular motion 39

   4.4   Angular velocity 40

   4.5   Angular acceleration 40

   4.6   Total acceleration 41

   4.7   Frequency 41

   4.8   Period 42

   4.9   Number of laps 42

   4.10   Center of rotation 43

   4.11   Distance 43

4.12 Arc length 44
4.13 Coordinate 44
4.14 Relative velocity 45

5 Examples circular kinematics of particles . . . 46

5.1 Example 1 46
5.2 Example 2 48
5.3 Example 3 50
5.4 Example 4 52
5.5 Example 5 54

6 Module statics of particles 57

6.1 Stationary reference system 58
6.2 Block 58
6.3 Block above an inclined plane to the left 59
6.4 Block above an inclined plane to the right 59
6.5 Springs 60
6.6 Pulley 60
6.7 Static point 61
6.8 Angles 61
6.9 Forces 62
6.10 Frictions 62
6.11 Vertical/Horizontal resultant 63
6.12 Resultant 63

7 Examples statics 64

7.1 Example 1 64
7.2 Example 2 66
7.3 Example 3 68
7.4 Example 4 70
7.5 Example 5 72
7.6 Example 6 74
7.7 Example 7 76
7.8 Example 8 78
7.9 Example 9 81

8   Module rigid statics 83

     8.1   Stationary reference system 84
     8.2   Points 84
     8.3   Beam 85
     8.4   Solid 86
     8.5   Point 87
     8.6   Angles 87
     8.7   Elements of beam 88
     8.8   Elements of solid 89
     8.9   Forces 90
     8.10   Frictions 90
     8.11   Couple 91
     8.12   Beams of 2 forces 92
     8.13   Truss 93
     8.14   Joint 93
     8.15   Beams of truss 94
     8.16   Resultant 95
     8.17   Resultant with horizontal force 95
     8.18   Resultant with vertical force 96

9   Examples statics rigid bodies 97

     9.1   Example 1 97
     9.2   Example 2 100
     9.3   Example 3 102
     9.4   Example 4 106
     9.5   Example 5 108
     9.6   Example 6 111
     9.7   Example 7 113
     9.8   Example 8 115
     9.9   Example 9 117
     9.10   Example 10 120
     9.11   Example 11 125
     9.12   Example 12 127
     9.13   Example 13 131

10   Module dynamics of particles 135

   10.1   Stationary reference system 137
   10.2   Mobile 138
   10.3   Mobile in X/Y 139
   10.4   Block with vertical movement 140
   10.5   Block with horizontal movement 141
   10.6   Block with movement along an inclined plane to the left 142
   10.7   Block with movement along an inclined plane to the right ... 143
   10.8   Pulley 144
   10.9   Springs 144
   10.10   Forces 145
   10.11   Frictions 145
   10.12   Frictions between blocks (contacts) 146
   10.13   Relation between accelerations 146
   10.14   Relative motion 147
   10.15   Collision 148
   10.16   Energy 149
   10.17   Momentum 150
   10.18   Power 150

11   Examples dynamics of particles 151

   11.1   Example 1 151
   11.2   Example 2 153
   11.3   Example 3 155
   11.4   Example 4 158
   11.5   Example 5 162
   11.6   Example 6 165
   11.7   Example 7 167
   11.8   Example 8 169
   11.9   Example 9 172
   11.10   Example 10 174
   11.11   Example 11 176
   11.12   Example 12 180

12   Module circular dynamics of particles 182

   12.1   Stationary reference system 185
   12.2   Object in rest 185
   12.3   Mobile with linear movement 186
   12.4   Mobile with circular movement 187
   12.5   Mobile with polar circular movement 188
   12.6   Mobile with perpendicular circular movement 189
   12.7   Angular velocity 190
   12.8   Centripetal acceleration 190
   12.9   Angular acceleration 191
   12.10   Energy 191

v

12.11 Angular momentum 192
12.12 Linear momentum 193
12.13 Power 194
12.14 Initial System 195
12.15 Final System 196
12.16 Center of rotation 197
12.17 Springs 198
12.18 Forces 198
12.19 Frictions.. 199
12.20 Angles.. 199
12.21 Moment of a force or couple of forces...................... 200
12.22 Total acceleration (Triangle of accelerations) 201
12.23 Maximum acceleration 201
12.24 Inertia.. 202
12.25 Absolute velocity 203
12.26 Sine of angle... 204
12.27 Supported combinations of elements Energy, Angular
Momentum, Linear Momentum and Power 205

13 Examples circular dynamics of particles .. 210
13.1 Example 1.. 210
13.2 Example 2.. 213
13.3 Example 3.. 216
13.4 Example 4.. 219
13.5 Example 5.. 224
13.6 Example 6.. 228
13.7 Example 7.. 233
13.8 Example 8.. 237
13.9 Example 9.. 242
13.10 Example 10... 246
13.11 Example 11... 251
13.12 Example 12... 253
13.13 Example 13... 256

14 Calorimetry 258
14.1 Laboratory clock.. 260
14.2 Applied heat.. 260
14.3 Applied heat flow 260
14.4 Heat extracted.. 261
14.5 Refrigeration 261
14.6 Block... 262
14.7 Liquid 262
14.8 Gas... 263
14.9 Linear expansion.. 263
14.10 Superficial expansion 264

14.11 Volumetric expansion 264
14.12 Change of state solid-liquid 265
14.13 Change of state liquid-gas............................... 266
14.14 Process.. 267
14.15 Calorimeter.. 268
14.16 Gas at constant pressure................................. 269
14.17 Gas at constant temperature............................. 269
14.18 Gas at constant volume.................................. 270
14.19 Gas PV/T.. 270
14.20 Heat exchanger 271

15 Examples calorimetry 272
15.1 Example 1.. 272
15.2 Example 2.. 273
15.3 Example 3.. 275
15.4 Example 4.. 276
15.5 Example 5.. 277
15.6 Example 6.. 278
15.7 Example 7.. 280
15.8 Example 8.. 282
15.9 Example 9.. 283
15.10 Example 10.. 284
15.11 Example 11.. 286
15.12 Example 12.. 288
15.13 Example 13.. 293

16 GNU Free Documentation License 294

Indice 303

1 Introduction

FisicaLab is an educational application to solve physics problems. Its main objective is let the user to focus in physics concepts, leaving aside the mathematical details (FisicaLab take care of them). This allows the user to become familiar with the physical concepts without running the risk of getting lost in mathematical details. And so, when the user gain confidence in applying physical concepts, will be better prepared to solve the problems by hand (with pen and paper). FisicaLab is easy to use and very intuitive. However, in order to take advantage of all its features, we recommend you read first these help files.

FisicaLab display two windows, one named *Modules and elements* and other named *Chalkboard*. The first of these windows, contain all modules that can be used to solve problems. These modules are grouped by: kinematics, static, dynamics, ... (see image below). You can select one of these groups with the buttons at the top of the window, marked with (1) in the image. When you leave the mouse's cursor above one of these buttons, a label with the group name is displayed. The buttons marked with (2) let you select the system of units, SI or English. You can see the modules of the selected group inside the box marked with (3). The tabs marked with (4) let you select one of the available modules. The elements of the selected module are displayed inside the box marked with (5). This elements let you set the problems. Inside the box marked with (6) you can write the element's data (if any element is selected, this box will be empty).

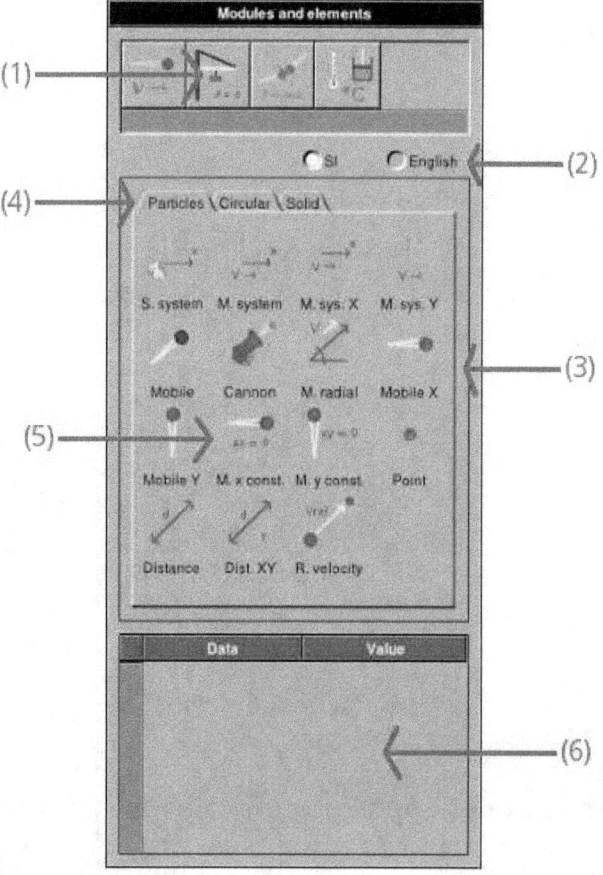

The window named *Chalkboard* (see image below), have at the top two buttons, marked with (7). The button at the left let you solve the problem, and the other is to clean the chalkboard. The black box marked with (8) is the chalkboard, where you add the elements to set the problems. You need keep in mind, although you can't see, that the chalkboard is a grid formed with cells of 50x50 pixels. By default the chalkboard size is 26x18 cells. In *Preferences* panel you can change the size to a maximum of 100x100 cells (A greater size than the default could be useful for trusses problems). The text view marked with (9) is where FisicaLab show the answer and messages. The checkbox marked with (10) erase the content of the text view before show the next answer or message. If you want keep the previous content, unselect this checkbox. In this case you can add notes to identify the results of the different problems.

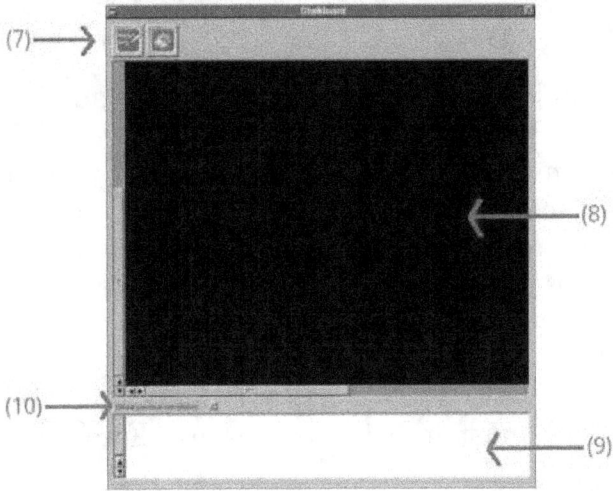

1.1 Handling the elements

To add elements at the chalkboard, do a click above the desired element. The mouse's cursor will become in an open hand, meaning this that we will add an element. Do a click above the chalkboard in the position where you want the element, the mouse's cursor will back at its original shape. Each time you add a new element, or select one different, a yellow square will be drawn around the current element. The data of the current element are displayed, for its edition, at panel *Modules and elements*. When you leave the mouse's cursor above one element in the chalkboard, a label with the element's data is displayed. In *Preferences* panel you can configure the font size of these labels.

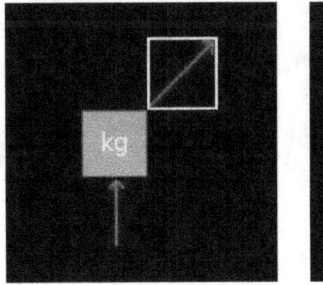 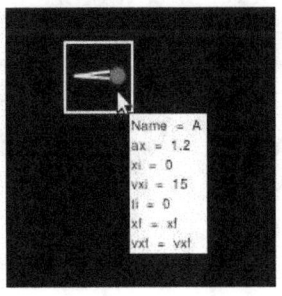

If you want move an element, click above it with the CONTROL key pressed, the mouse's cursor will become in a close hand, meaning this that we are moving an element. Do click in the new position and the mouse's cursor will back at its original shape. In other hand, if you want delete an element, do click above it with the SHIFT key pressed.

Keep in mind that FisicaLab don't let you combine elements from different modules. The elements in each module are enough to set a wide variety of problems.

1.2 Element data

When you select an element in the chalkboard, or add a new element, you will see a table at the bottom of the window *Modules and elements*. With a double click above any field of the second column, you can write the data. FisicaLab supports scientific notation, to use this use the letter E. For example, to write the number '3.45x10-5', write:

```
3.45E-5
```

All numerical data must be without spaces. For example, the following numbers are wrong:

```
- 5.3
7.8E - 8
```

Also, FisicaLab can use many conversion factors. To use these, add the character @ before the conversion. If you have selected the SI system, FisicaLab assumes that all data are in meters, kg, seconds, etc. With the English system, FisicaLab assumes that all data are in feet, pounds, slugs, seconds, etc. (in the English system the mass must be in slugs). For example, if you want write an speed of '75 km/h', use:

```
75 @ km/h
```

Here, we have one space before and after the character @, but these are for clarity, and are not required. Each module has its how conversion factors, as you can see in the sections that deal about these.

You can use letters or words to represent the unknown data. If, for example, the final velocity is an unknown data, you can represent this like:

```
fv
finalv
fvel
```

or any other combination. But, we recommend you use letters or words that are related with the unknown data. Also, the conversion factors can be used with the unknown data. For example, if the time is unknown, and you want this in minutes, write something like:

```
t @ min
```

The scientific notation can be used with the unknowns, adding the characters *#E* at the end of the name. For example, for a coefficient of thermal expansion, that is a small value:

```
coefficient#E
```

Also, this can be used with a conversion factor. For example, for a very long distance that we want in kilometers:

```
distance#E @ km
```

All the conversion factors are available in a contextual menu. After select the row of data where want add the factor, a right mouse click open a context menu with all available factors.

FisicaLab allows mathematical operations directly on the fields where you enter data. Can be carried out operations of addition (+), subtraction (-), multiplication (*) and division (/). Although not allowed to group operations by parentheses. It also provides some useful functions for certain calculations. These are listed below with its description:

cos(ang)   Calculates the cosine of the sexagesimal angle "ang".

sin(ang)   Calculates the sine of the sexagesimal angle "ang".

tan(ang)   Calculates the tangent of the sexagesimal angle "ang".

sqrt(x)   Calculates the square root of the number "x".

hypot(a,b)   Calculates the hypotenuse of a right triangle whose legs are "a" and "b".

leg(c,a)   Calculates the leg of the right triangle whose hypotenuse is "c" and the other leg is "a".

rd(m1,m2,d)   Calculates the distance of the mass "m1" to the center of mass of the system consisting of the masses "m1" and "m2", which are spaced a distance "d".

The numbers that are passed as parameters to these functions must have consistent units. For example, in the 'hypot()' function both legs must be in the same units, whether centimeters, meters, inches, etc. These functions can be used in operations of addition, subtraction, multiplication and division. In these operations blank spaces are not allowed. Here are some examples:

```
8*cos(34)
hypot(4,3)-2
rd(3,6,40)*sin(30) @ cm
15*8/hypot(13,8)
```

Once entered the operation, FisicaLab will do the calculation and will write the result in the entry. Note that is possible to apply conversion factors. Although these can also be applied after carrying out the calculation.

The fields where you enter angles do not allow the operations and functions described above. This is because these fields have their own operations and functions. For example, FisicaLab allows write the angles as slopes (a/b), and automatically convert this to sexagesimal angles. What is very useful for problems of trusses.

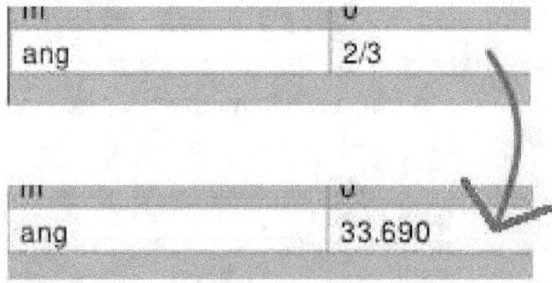

For these fields are available two functions that calculates an angle from other relationships. These functions are:

acos(a/c)   Calculates the sexagesimal angle whose co-
                             sine is the ratio "a/c".

asin(b/c)   Calculates the sexagesimal angle whose sine
                             is the ratio "b/c".

If an operation is entered incorrectly, for example if it contains spaces or contains a function with an incorrect number of parameters, FisicaLab will do nothing and will take that string as an unknown.

Caution: *If, for example, you add a mass conversion factor in a time data, this will cause an error in the solution. And you will not get a message about this error.*

1.3 How it works

FisicaLab work over the base of *number of equations = number of unknown data*. In general you don't need worry about this. But in some cases you will see the error **"The system is undetermined"**. This occurs when you write numeric data in a field that must be an unknown data. The examples show this cases.

1.4 Messages

FisicaLab write a wide variety of messages in the text view when a problem is wrong. However, you always will see a message about the calculation's status, as you can see in the following image:

The last line say **"State = success"**, meaning that the calculation was successful. Any other status different as *success*, mean that or the set problem don't have a solution, or an unexpected error occurred.

2 Module kinematics of particles

The conversion factors for SI system are:

km	kilometer
cm	centimeter
mm	millimeter
mi	mile
ft	feet
in	inch
h	hour
min	minute
km/h	kilometers per hour
cm/s	centimeters per second
mm/s	millimeters per second
mph	miles per hour
ft/s	feet per second
in/s	inch per second
kt	knot
cm/s2	centimeters per squared second
mm/s2	millimeters per squared second
ft/s2	feet per squared second
in/s2	inch per squared second

And for the English system are:

km	kilometer
m	meter
cm	centimeter
mm	millimeter
mi	mile
in	inch
h	hour
min	minute
km/h	kilometers per hour
m/s	meters per second
cm/s	centimeters per second
mm/s	millimeters per second
mph	miles per hour
in/s	inch per second
kt	knot
m/s2	meters per squared second
cm/s2	centimeters per squared second
mm/s2	millimeters per squared second
in/s2	inch per squared second

This module have 15 elements, presented below. With a description of each one, its data and its number of equations.

2.1 Stationary reference system

An stationary reference system, with the X axis horizontal and positive to the right, and the Y axis vertical and positive to upwards.

Equations: None.

Data:
t: End time of the problem.

2.2 Mobile reference system

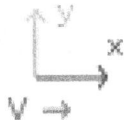

A mobile reference system.

Equations: 4

Data:

Name: Name of the mobile system.

Object: Name of the object inside the mobile system.

xsi: Initial X coordinate of the mobile system in reference to the stationary system.

ysi: Initial Y coordinate of the mobile system in reference to the stationary system.

vsx: Velocity in X of the mobile system.

vsy: Velocity in Y of the mobile system.

xof: Final X coordinate of the object in reference to the stationary system.

yof: Final Y coordinate of the object in reference to the stationary system.

vxof: Final velocity in X of the object in reference to the stationary system.

vyof: Final velocity in Y of the object in reference to the stationary system.

2.3 Mobile reference system in X/Y

Mobile reference systems in X and Y, respectively.

Equations: 2

Data:

Name: Name of the mobile system.

Object: Name of the object inside the mobile system.

xsi (ysi):   Initial X (Y) coordinate of the mobile system in reference to the stationary system.

vsx (vsy):   Velocity in X (Y) of the mobile system.

xof (yof):   Final X (Y) coordinate of the object in reference to the stationary system.

vxof (vyof):   Final velocity in X (Y) of the object in reference to the stationary system.

2.4 Mobile

A general mobile.

Equations: 4

Data:

Name: Name of the mobile.

ax: Acceleration in X.

ay: Acceleration in Y.

xi: Initial coordinate in X.

yi: Initial coordinate in Y.

vxi: Initial velocity in X.

vyi: Initial velocity in Y.

ti: Time at which movement begins.

xf: Final X coordinate.

yf: Final Y coordinate.

vxf: Final velocity in X.

vyf: Final velocity in Y.

2.5 Mobile in X/Y

Mobiles in X and Y respectively.

Equations: 2

Data:

Name: Name of the mobile.

ax (ay):   Acceleration in X (Y).

xi (yi):   Initial coordinate in X (Y).

vxi (vyi):   Initial velocity in X (Y).

ti: Time at which movement begins.

xf (yf):   Final X (Y) coordinate.

vxf (vyf):   Final X (Y) coordinate.

2.6 Mobile in X/Y with constant velocity

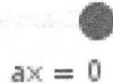

ax = 0

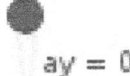

ay = 0

Mobiles in X and Y with constant velocity, respectively.

Equations: 1

Data:

Name: Name of the mobile.

xi (yi):   Initial X (Y) coordinate.

xf (yf):   Final X (Y) coordinate.

ti: Time at which movement begins.

vx (vy):   Constant velocity in X (Y).

2.7 Cannon

Cannon to simulate parabolic shots. The angles are measured from the positive X axis, and the positive sense is the opposite of clockwise.

Equations: 4

Data:

Name:	Name of the cannon.
ax:	Acceleration in X axis.
ay:	Acceleration in Y axis.
xi:	Initial X coordinate.
yi:	Initial Y coordinate.
vi:	Initial velocity.
angi:	Angle of the launch.
ti:	Time at which movement begins.
xf:	Final X coordinate.
yf:	Final Y coordinate.
vf:	Final velocity.
angf:	Angle of the final velocity vector.

2.8 Mobile radial

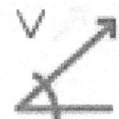

Mobile radial with velocity and acceleration along the direction specify by the angle. The angle is measured from the positive X axis, and the positive sense is the opposite of the clockwise.

Equations: 3

Data:

Name:	Name of the mobile.
a:	Acceleration.
angf:	Angle that specify the direction.
xi:	Initial X coordinate.
yi:	Initial Y coordinate.
vi:	Initial velocity.
ti:	Time at which movement begins.
xf:	Final X coordinate.
yf:	Final Y coordinate.
vf:	Final velocity.

2.9 Distance

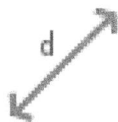

To measured the distance between mobiles, at end time. Or between a mobile and one point.

Equations: 1

Data:

Object 1: Name of the first object.

Object 2: Name of the second object.

d: Distance between the objects.

2.10 Distance XY

To measured the distance between objects along one axis (X or Y).

Equations: 1

Data:

x1 (y1): X or Y coordinate of the first object.

x2 (y2): X or Y coordinate of the second object.

x1 - x2 (y1 - y2): Difference between x1 and x2 (or between y1 and y2).

2.11 Point

Point of reference to measured a distance.

Equations: None.

Data:

xf: X coordinate of the point.

yf: Y coordinate of the point.

2.12 Relative velocity

Measured the vector relative velocity (magnitude and angle) of the Object 1 relative to Object 2. The angles are measured from the positive X axis, and the positive sense is the opposite of the clockwise.

Equations: 2

Data:

Object 1: Object that relative velocity you want.

Object 2: Object reference to measured the relative velocity.

v: Magnitude of the vector relative velocity.

ang: Angle of the vector relative velocity.

3 Examples kinematics of particles

3.1 Example 1

A car start from the rest with an acceleration of 3 m/s2. Which velocity, in km/h, have after 7 seconds? Which is the travelled distance in meters?

Solution with FisicaLab

Select the Kinematic group and, inside this, the Particles module. Erase the content of the chalkboard. And add one element Mobile in X and one element Stationary reference system. As show the image below:

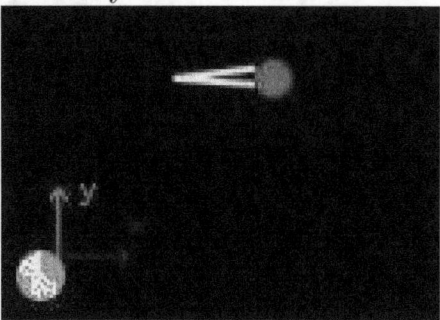

We know the end time, 7 seconds. Then in the element Stationary reference system write this.

tf   7

To the element Mobile in X, we assume that the car begins its movement from the X coordinate 0 and at the time 0. Also, we assume that its movement is in the direction of the positive X axis. But, as we want the final velocity in km/h, we need write:

    vf @ km/h

To the final velocity. As show below:

Name	Car
ax	3
xi	0
vxi	0
ti	0

xf   d

vxf   vf @ km/h

Where *d* is the final X coordinate, an unknown data. To this example, the name of the mobile *Car* is irrelevant. Now click in the icon Solve to get the answer.

```
d = 73.500 m ;   vf = 75.600 km/h ;
Status = success.
```

3.2 Example 2

A car start from a place B with a constant velocity of 75 km/h at the east. 8 minutes after, other car with a constant velocity of 88 km/h start from a place A to the east. The place A is 5 km to the west of B. How minutes needs the last car to reach the first car? In that moment, which is the distance between the place A and the cars?

Solution with FisicaLab

Select the Kinematic group and, inside this, the Particles module. Erase the content of the chalkboard. And add two elements Mobile in X with constant velocity and one element Stationary reference system. As show the image below:

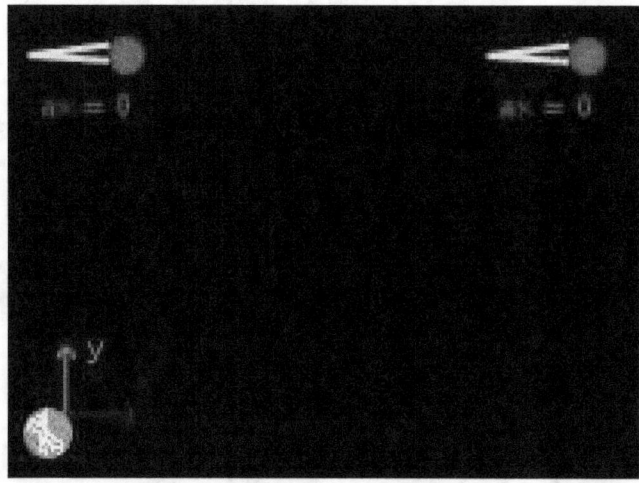

We assume that the Mobile in X at the left, corresponds to the car that start from A. Now, the end time, is an unknown data, then in the element Stationary reference system write:

tf   t @ min

Because we want the end time in minutes. Now, we assume that the car that start from A is in the origin of the stationary reference system, and start at 8 minutes. We call this car *Car A*:

Name	Car A
xi	0
xf	d @ km
ti	8 @ min
vx	88 @ km/h

To the car that start from B. Its initial X coordinate is 5 km (because the other car is in the origin), and the start time is 0. We call this car *Car B*:

Name   Car B

xi   5 @ km

xf   d @ km

ti   0

vx   75 @ km/h

The final X coordinate of both cars is the same, because we want know the moment when the last car reach the first car. Now click in the icon Solve to get the answer.

```
d = 101.538 k m ;   t = 77.231 min ;
Status = success.
```

3.3 Example 3

Two cars start from the same place. One with a constant velocity of 80 km/h at 50 degrees from the east, in the direction north-east. The other with a constant velocity of 95 km/h to the east. In what moment the distance between these cars is 50 km? Give your answer in minutes.

Solution with FisicaLab

Select the Kinematic group and, inside this, the Particles module. Erase the content of the chalkboard. And add one element Mobile in X whit constant velocity, one element Mobile radial, one element Distance and one element Stationary reference system. As show the image below:

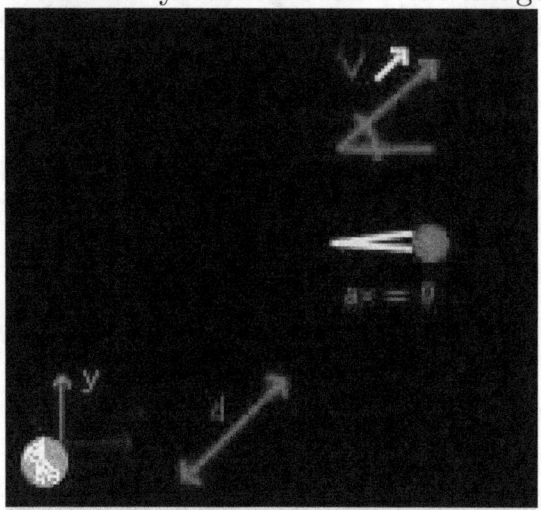

The end time, is an unknown data, then in the element Stationary reference system write:

tf   t @ min

Because we want the time in minutes. Now, we assume that both cars start from the origin of the stationary reference system, and at the time 0. Then, to the element Mobile radial write, we call this car *Car 1*:

Name	Car 1
a	0
angf	50
xi	0
yi	0
vi	88 @ km/h
ti	0

xf	xf1
yf	yf1
vf	vf1

We write the final velocity as an unknown data. Although, the acceleration is 0. We do this to satisfied *numbers of equations = numbers of unknown data*. Now, to the Mobile in X write, we call this *Car 2*:

Name	Car 2
xi	0
xf	xf2
ti	0
vx	95 @ km/h

To the element Distance:

Object 1	Car 1
Object 2	Car 2
d	50 @ km

Where Object 1 and Object 2 are, respectively, the *Car 1* and *Car 2*. And *d* is the distance between both cars, in this case 50 kilometers.

Now click in the icon Solve to get the answer.

```
xf1 = 36447.231 m ;  yf1 = 43436.119 m ;
vf1 = 24.444 m/s ;  t = 38.660 min ;
xf2 = 61212.198 m ;
Status = success.
```

The desired data is t = 38.660 min.

3.4 Example 4

A cannon shot a projectile at 35 degrees from the horizontal, with a velocity of 17 m/s. We want know the maximum altitude, and the range.

Solution with FisicaLab

Select the Kinematic group and, inside this, the Particles module. Erase the content of the chalkboard. And add one element Cannon and one element Stationary reference system. As show the image below:

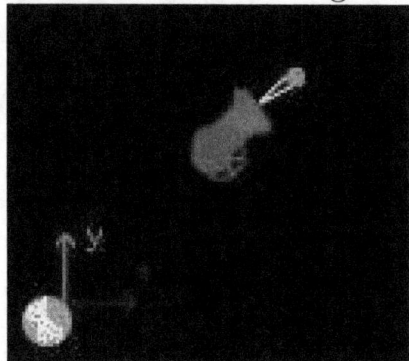

The end time, is an unknown data, then in the element Stationary reference system write:

tf   t

The data of the maximum altitude and range correspond at different times. Then, we need first set the problem to determine the maximum altitude, and after set the problem to determine the range.

Maximum altitude

To this case, to the element Cannon write:

Name	Projectile
ax	0
ay	-9.81
xi	0
yi	0
vi	17
angi	35
ti	0

xf	xf
yf	yf
vf	vf
angf	0

The acceleration in X axis is 0, but in Y is -9.81 (the acceleration of gravity). We assume that the cannon shot at time 0. And that this is located at the origin of the stationary system. Also, we know that, at the maximum altitude, the angle of the velocity vector is 0.

Now click in the icon Solve to get the answer.

```
xf = 13.842 m ;   yf = 4.846 m ;   vf = 13.926 m/s ;
t = 0.994 s ;
Status = success.
```

The desired data is yf = 4.846 m.

Range

There are many ways to calculate the range. Here, we write to the end Y coordinate 0, and to the final velocity and the end angle write unknown data. This is necessary to satisfied numbers of equations = numbers of unknown data. Then, write:

Name	Projectile
ax	0
ay	-9.81
xi	0
yi	0
vi	17
angi	35
ti	0
xf	xf
yf	0
vf	vf
angf	angf

Now click in the icon Solve to get the answer:

```
xf = 27.683 m ;   vf = 17.000 m/s ;   angf = -35.000 degrees ;
t = 1.988 s ;
Status = success.
```

NOTE: *If you want the altitude in other position, keep in mind that, in general, are two solutions. The first when the projectile is ascending and the second when the projectile is descending. If you want the second position, you need set the initial position (the start position) after the first position.*

3.5 Example 5

A swimmer want to cross a river with a width of 25 meters. Its current have a velocity of 7 m/s. The swimmer take a perpendicular direction to the rive. If the swimmer have a velocity of 7 m/s, how minutes needs to cross the river?, what distance was dragged the swimmer?

Solution with FisicaLab

The velocity of the swimmer is in reference to the river, that is a mobile reference system in reference to the share. With this in mind, erase the content of the chalkboard. And add one element Mobile reference system in X, one element Mobile in Y with constant velocity, and one element Stationary reference system. As show the image below:

We assume that the river's movement is in the direction of the positive X axis. That the swimmer take the direction of the positive Y axis. And that both, the reference system and the swimmer, start from the origin of the stationary reference system. The end time is an unknown data:

tf   t

To the Mobile in Y with constant velocity, we call this *Swimmer*, write:

Name	Swimmer
yi	0
yf	25
ti	0
vy	2

To the Mobile reference system in X, write:

Name	Mobile system
Object	Swimmer

xsi	0
vsx	7
xof	xf
vxof	vf

We write the final velocity to the swimmer in reference to the stationary system, vxof, as an unknown data, to satisfied *numbers of equations = numbers of unknown data*. Of course, this velocity is 7 m/s, because the swimmer don't have a velocity in X axis. Now click in the icon Solve to get the answer.

```
t = 12.500 s ;  xf = 87.500 m ;  vf = 7.000 m/s ;
Status = success.
```

3.6 Example 6

Two cars, A and B, start from the same place. A take the direction of the north with a constant velocity of 80 km/h, and B the direction of the east, with an acceleration of 4 m/s2 (start from the rest). After 13 seconds, which is the relative velocity of the car A in reference to B?

Solution with FisicaLab

Erase the content of the chalkboard. And add one element Mobile in X, one element Mobile in Y with constant velocity, one element Relative velocity, and one element Stationary reference system. As show the image below:

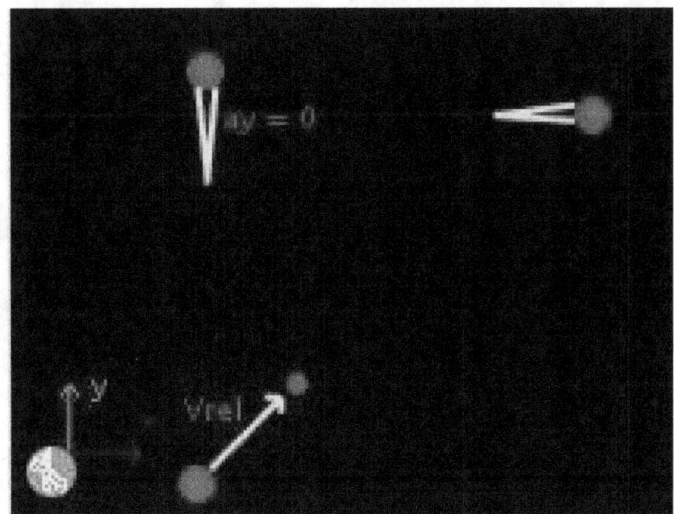

We assume that the car A take the direction of the positive Y axis, and car B the direction of the positive X axis. And that both cars start from the origin. Now, as we know the end time, in the element Stationary reference system, write:

tf   13

To the element Mobile in Y with constant velocity, we call this car *Car A*, write:

Name   Car A
yi   0
yf   yf
ti   0
vy   80 @ km/h

To the element Mobile in X, we call this car *Car B*:

Name	Car B
ax	4
xi	0
vxi	0
ti	0
xf	xf
vxf	vxf

And to the element Relative velocity:

Object 1	Car A
Object 2	Car B
v	v
ang	ang

Where the magnitude and the angle of the relative velocity are unknown data. Now click in the icon Solve to get the answer:

```
v = 56.549 m/s ;   ang = 156.861 degrees ;   xf = 338.000 m ;
vxf = 52.000 m/s ;   yf = 288.889 m ;
Status = success.
```

The desired data are v = 56.549 m/s and ang = 156.861 degrees.

3.7 Example 7

A car A travel with a velocity of 110 km/h, in some moment pass together a B car, that is in rest. 4 seconds after, the car B start with an acceleration of 2.6 m/s2 in chase to the car A. How seconds need to be 5 meters behind the car A?

Solution with FisicaLab

Erase the content of the chalkboard. And add one element Mobile in X, one element Mobile in X with constant velocity, one element Distance XY, and one element Stationary reference system. As show the image below:

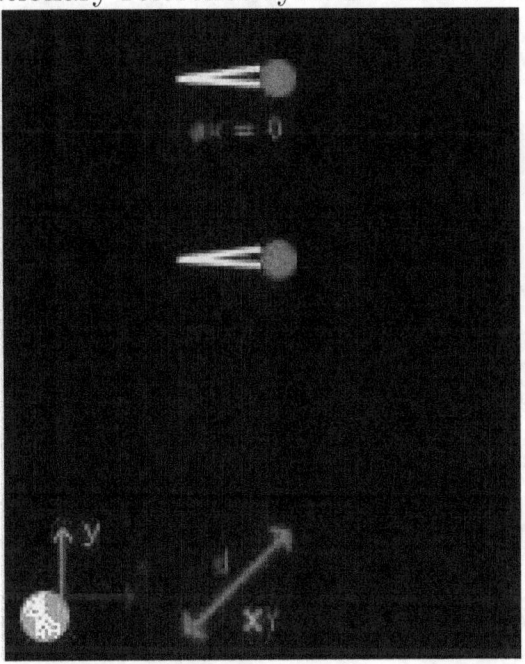

The end time is an unknown data:

tf   t

We assume that the car A is moving in the direction of the positive X axis. And that both cars start from the origin. The car A is the element Mobile in X with constant velocity. To this, write:

Name	Car A
xi	0
xf	xfA
ti	0
vx	110 @ km/h

To the element Mobile in X, the car B, write (remember that this car begins its movement after 4 seconds):

Name	Car B
ax	2.6
xi	0
vxi	0
ti	4
xf	xfB
vxf	vfB

To the element Distance XY:

x1 (y1)	xfA
x2 (y2)	xfB
x1 - x2 **(y1 - y2)**	5

Now click in the icon Solve to get the answer.

```
t = 30.861 s ;  xfA = 942.981 m ;  xfB 937.981 ;
vfB = 69.839 m/s ;
Status = success.
```

4 Module circular kinematics of particles

The conversion factors for SI system are:

km	kilometer
cm	centimeter
mm	millimeter
mi	miles
ft	feet
in	inch
h	hour
min	minute
km/h	kilometer per hour
cm/s	centimeter per second
mm/s	millimeter per second
mph	miles per hour
ft/s	feet per second
in/s	inch per second
kt	knot
cm/s2	centimeters per squared second
mm/s2	millimeters per squared second
ft/s2	feet per squared second
in/s2	inch per squared second
rad	radian
rpm	revolutions per minute

And for the English system are:

km	kilometer
m	meter
cm	centimeter
mm	millimeter
mi	miles
in	inch
h	hour
min	minute
km/h	kilometers per hour
m/s	meters per second
cm/s	centimeters per second
mm/s	millimeters per second
mph	miles per hour
in/s	inch per second
kt	knot
m/s2	meters per squared second
cm/s2	centimeters per squared second
mm/s2	millimeters per squared second
in/s2	inch per squared second
rad	radian
rpm	revolutions per minute

This module have 14 elements, presented below. With a description of each one, its data and its number of equations.

4.1 Stationary reference system

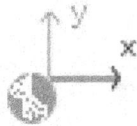

An stationary reference system, with the X axis horizontal and positive to the right, and the Y axis vertical and positive to upwards.

Equations: None.

Data:
t: End time of the problem.

4.2 Mobile with circular motion

Mobile with circular motion (constant radius). The angles are measured from the positive X axis, the positive sense is the opposite of clockwise. Also, a positive value of the tangential velocity, mean that the mobile tour in the opposite sense of clockwise.

Equations: 4

Data:

Name: Name of the mobile.

C: Point that makes the center of rotation. If
               not specified the origin is the center.

r: Radius of circular motion.

aci: Centripetal acceleration at initial time.

at: Tangential acceleration.

angi: Angle of the initial position.

vi: Initial tangential velocity.

ti: Time at which movements begins.

angf: Angle of the final position.

vf: Final tangential velocity.

acf: Centripetal acceleration at end time.

4.3 Mobile with polar circular motion

   Mobile with circular motion described by polar coordinates. The angles are measured from the X axis, the positive sense is the opposite of clockwise. Also, a positive value to angular velocity, mean that the mobile tour in the opposite sense of clockwise. And a positive value to radial velocity, indicates that the mobile moves away from the center of rotation.

Equations: 4

Data:

Name: Name of the mobile.

C: Point that makes the center of rotation. If
               not specified the origin is the center.

aa: Angular acceleration.

ar: Radial Acceleration.

angi: Angle of the initial position.

ri: Initial radius of the mobile.

vai: Initial angular velocity.

vri: Initial radial velocity.

ti: Time at which movements begins.

angf: Angle of the final position.

rf: Final radius of the mobile.

vaf: Final angular velocity.

vrf: Final radial velocity.

4.4 Angular velocity

Measure the angular velocity of the mobile indicated. Only applies to elements Mobile with circular motion.

Equations: 2

Data:

Object: Name of the mobile.

vangi: Angular velocity at initial time.

vangf: Angular velocity at end time.

4.5 Angular acceleration

Measure the angular acceleration of the mobile indicated. Only applies to elements Mobile with circular motion.

Equations: 1

Data:

Object: Name of the mobile.

aang: Angular acceleration.

4.6 Total acceleration

Measure the total acceleration of the mobile indicated. The angle is measured from the radius vector, the positive sense is the opposite of clockwise.

Equations: 4

Data:

Object: Name of the mobile.

atoti: Magnitude of total acceleration at initial time.

aangi: The angle of the initial total acceleration vector.

atotf: Magnitude of total acceleration at end time.

aangf: The angle of the end total acceleration vector.

4.7 Frequency

Measure the frequency of the mobile indicated. If the mobile have a tangential o radial acceleration, the measured frequency is the average value.

Equations: 1

Data:

Object: Name of the mobile.

f: Frequency.

4.8 Period

T

Measure the period of the mobile indicated. If the mobile have a tangential o radial acceleration, the measured period is the average value.

Equations: 1

Data:
Object: Name of the mobile.

T: Period.

4.9 Number of laps

Measure the number of laps of the mobile indicated.

Equations: 1

Data:
Object: Name of the mobile.

n: Number of laps.

4.10 Center of rotation

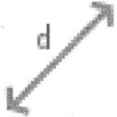

Especifica un centro de rotación.

Equations: None

Data:
Name: Name of the point.

x: X coordinate of the point.

y: Y coordinate of the point.

4.11 Distance

Measure the distance between two mobiles at final time.

Equations: 1

Data:
Object 1:   Name of the first object.

Object 2:   Name of the second object.

d: Distance between the mobiles.

4.12 Arc length

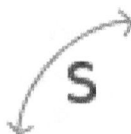

Measure the length of the arc described by the mobile indicated. If the mobile have a radial motion, the measured value is approximate.

Equations: 1

Data:

Object: Name of the mobile.

s: Arc length.

4.13 Coordinate

Get the coordinate of the specified mobile at the end time.

Equations: 2

Data:

Object: Name of the mobile.

x: X coordinate.

y: Y coordinate.

4.14 Relative velocity

 Vrel

Measure the relative velocity of mobile Object 1 relative to mobile Object 2. The angle is measured from the positive X axis, the positive sense is the opposite of clockwise. Also, this element can be used to measure the relative velocity (in this case the total velocity) of a Mobile with polar circular motion relative to its center of rotation.

Equations: 2

Data:

Object 1:   Name of first mobile.

Object 2:   Name of second mobile.

v: Magnitude of the relative velocity.

ang: Angle of the relative velocity vector.

5 Examples circular kinematics of particles

5.1 Example 1

A body in uniform circular motion goes through a circle of radius 3 m with a constant tangential velocity of 27 m/s. What is the centripetal acceleration of the mobile? What is the distance traveled after 6.3 seconds (in meters)?

Solution with FisicaLab

Select the Kinematic group and, inside this, the Particles circular module. Erase the content of the chalkboard and select the SI system. Now add one element Mobile with circular motion, one element Arc length and one element Stationary reference system. As show the image below:

To the element Stationary reference system we have:

tf   6.3

The element Mobile with circular motion gonna be called *Mobile*. And we assume that the initial position correspond to the angle 0. Also, we left the center like 0 (to this problem the center of rotation is irrelevant). And the final velocity is left as unknown data, because the element have an entry to tangential acceleration and this assume that the final velocity is different from the initial. Also the centripetal acceleration at initial and end time are unknown data:

Name	Mobile
C	0
r	3
aci	aci

at	0
angi	0
vi	27
ti	0
angf	angf
vf	vf
acf	acf

In the element Arc length we set our *Mobile*, and the arc length value is of course an unknown data:

Object	Mobile
s	s

Now click in the icon Solve to get the answer:

```
aci = 243.000 m/s2 ;   angf = 8.671 degrees ;
vf = 27.000 m/s ;   acf = 243.000 m/s2 ;
s = 170.100 m ;
Status = success.
```

5.2 Example 2

A mobile left the rest describing a circular motion with radius 4.7 m. If the tangential acceleration is 1.6 m/s2, How many seconds need to travel a distance of 256 m? How many laps have done? What is the total acceleration in that moment (after travel the 256 m)?

Solution with FisicaLab

Select the Kinematic group and, inside this, the Particles circular module. Erase the content of the chalkboard and select the SI system. Now add one element Mobile with circular motion, one element Arc length, one element Number of laps, one element Total acceleration and one element Stationary reference system. As show the image below:

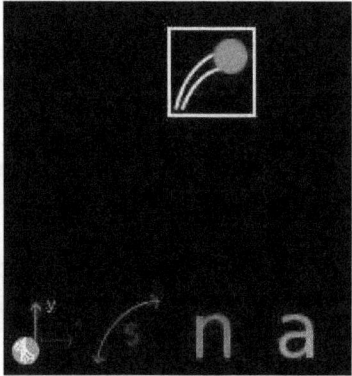

The time is unknown, then to the Stationary reference system we have:

tf   t

Now to the element Mobile with circular motion, we call it *Mobile*, we assume that its initial position correspond with angle 0. And, to this problem, we don't need a center of rotation:

Name	Mobile
C	0
r	4.7
aci	aci
at	1.6
angi	0
vi	0
ti	0
angf	angf

vf	vf
acf	acf

To the element Arc length we have:

Object	Mobile
s	256

Now to the element Number of laps:

Object	Mobile
n	n

And to the element Total acceleration:

Object	Mobile
atoti	atoti
angi	aangi
atotf	atotf
angf	aangf

Now click in the icon Solve to get the answer:

```
aci = -0.000 m/s2 ;  angf = 240.791 degrees ;
vf = 28.622 m/s ;   acf = 174.298 m/s2 ;
t = 17.889 s ;  n = 8.669 rev ;
atoti = 1.600 m/s2 ;  aangi = 90.000 degrees ;
atotf = 174.305 m/s2 ;  aangf = 179.474 degrees ;
Status = success.
```

Remember that the angle of the total acceleration vector is measured from the radius vector.

5.3 Example 3

A mobile with uniform circular motion describes a circle of radius 7.5 m. If the angular velocity is 1.3 rad/s, How seconds need to describe 9 laps? What is the centripetal acceleration? What is its period?

Solution with FisicaLab

Select the Kinematic group and, inside this, the Particles circular module. Erase the content of the chalkboard and select the SI system. Now add one element Mobile with circular motion, one element Angular velocity, one element Number of laps, one element Period and one element Stationary reference system. As show the image below:

The time is an unknown data, then to the element Stationary reference system:

tf	t

To the element Mobile with circular motion, gonna be called *Mobile*, the initial and final velocity are unknown data. Also the centripetal acceleration at initial and end time are unknown data:

Name	Mobile
C	0
r	7.5
aci	0
at	0
angi	0
vi	vi
ti	0
angf	angf

vf   vf

acf   acf

In the element Angular velocity, the value 1.3 rad/s, is the constant angular velocity of our *Mobile*, because this have a tangential acceleration equal to 0. However, to comply with *number of equations = number of unknown data*, we put this data as initial angular velocity and leave the final data as unknown, although it has the same value:

Object   Mobile

vangi   1.3

vangf   vangf

To the element Number of laps:

Object   Mobile

n   9

And to the element Period:

Object   Mobile

T   T

Now click in the icon Solve to get the answer:

```
aci = 12.675 m/s2 ;   vi = 9.750 m/s ;
angf = 360.000 degrees ;   vf = 9.750 m/s ;
acf = 12.675 m/s2 ;   t = 43.499 s ;
vangf = 1.300 rad/s ;   T = 4.833 1/hz ;
Status = success.
```

5.4 Example 4

A mobile left the rest from the origin of the coordinate system, with an angular velocity of 0.4 rad/s, a radial velocity of 0.2 m/s (in the direction of the positive X axis) and with a radial acceleration of 0.12 m/s2. What is the coordinate of its position after 5 seconds? What is its radial velocity at that time? What is its distance from the origin? What is the average frequency of its circular motion?

Solution with FisicaLab

Select the Kinematic group and, inside this, the Particles circular module. Erase the content of the chalkboard and select the SI system. Now add one element Mobile with polar circular motion, one element Coordinate, one element Frequency and one element Stationary reference system. As show the image below:

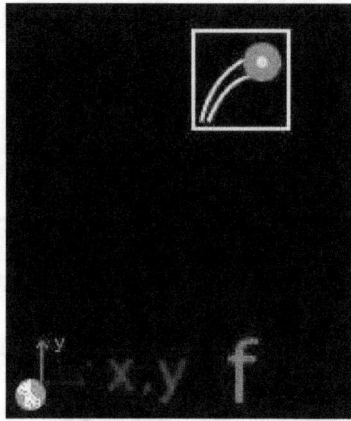

We know the time, then to the element Stationary reference system:

tf   5

Now to the element Mobile with polar circular motion, gonna be called *Mobile*, its initial position correspond with the angle 0 (because the initial radial velocity have the direction of the positive X axis). The center of rotation, is established as 0, because the mobile tour around the origin. The initial radius is 0 (because left the origin), the angular acceleration is 0 too, and the final angular velocity is established as an unknown data, because the element have an entry to angular acceleration then this assume that the final angular velocity can be different from the initial. And the final radial velocity is an unknown data:

Name	Mobile
C	0
aa	0

ar	0.12
angi	0
ri	0
vai	0.4
vri	0.2
ti	0
angf	angf
rf	rf
vaf	vaf

To the element Coordinate, we set our *Mobile* as the measured object, and the coordinates X and Y are unknown data:

Object	Mobile
x	x
y	y

Now in the element Frequency, the frequency is an unknown data too:

Object	Mobile
f	f

Now click in the icon Solve to get the answer:

```
x = -1.040 m ;   y = 2.273 m ;   f = 0.064 hz ;
angf = 114.592 degrees ;   rf = 2.500 m ;
vaf = 0.400 rad/s ;   vrf = 0.800 m/s ;
Status = success.
```

The distance of the mobile to the origin is, of course, the final radius 2.500 m.

5.5 Example 5

A mobile A with an uniform circular motion, tour around the origin describing a circle with radius 83 cm and with a tangential velocity of 0.7 m/s. In the moment in which the radius vector of this mobile is 59 degrees above the horizontal, a mobile B left the point (10, 17)cm with a constant radial velocity of 8 cm/s (in the direction of the positive X axis), and with a constant angular velocity of 0.6 rad/s. 12 seconds after the mobile B begin its movement, What is the distance between the two mobiles? What is the relative velocity of the mobile B in reference to the mobile A? How many laps described each of the mobiles?

Solution with FisicaLab

Select the Kinematic group and, inside this, the Particles circular module. Erase the content of the chalkboard and select the SI system. Now add one element Mobile with circular motion, one Mobile with polar circular motion, one element Center of rotation, one element Distance, one element Relative velocity, two elements Number of laps and one element Stationary reference system. As show the image below:

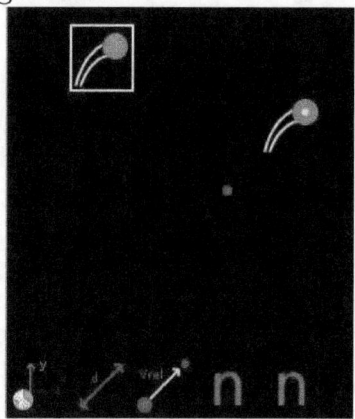

To the element Stationary reference system:

tf	12

To the element Mobile with circular motion (the mobile A that tour around the origin):

Name	A
C	0
r	83 @ cm
aci	aciA
at	0

angi	59
vi	0.7
ti	0
angf	angfA
vf	vfA
acf	acfA

To the element Center of rotation, the center of the mobile B, we have:

Name	Center
x	10 @ cm
y	17 @ cm

Now to the element Mobile with polar circular motion:

Name	B
C	Center
aa	0
ar	0
angi	0
ri	0
vai	0.6
vri	8 @ cm/s
ti	0
angf	angfB
rf	rfB
vaf	vafB
vrf	vrfB

The element Distance:

Object 1	A
Object 2	B
d	d

And the element Relative velocity:

Object 1 B
Object 2 A
v vrBA
ang angBA

And for each of the elements Number of laps, one for each mobile:

Object A
n nA

Object B
n nB

Now click in the icon Solve to get the answer:

```
aciA = 0.590 m/s2 ;   angfA = 278.861 degrees ;
vfA = 0.700 m/s ;   acfA = 0.590 m/s2 ;
d = 1.838 m ;   vrBA = 1.142 m/s ;
angBA = 164.452 degrees ;   nA = 1.611 rev ;
nB = 1.146 rev ;   angfB = 52.530 degrees ;
rfB = 0.960 m ;   vafB = 0.600 rad/s ;
vrfB = 0.080 m/s ;
Status = success.
```

6 Module statics of particles

The conversion factors for SI system are:

cm   centimeter

g   gram

slug   slug

T   metric ton

And for the English system are:

kg   kilogram

g   gram

in   inch

lb/in   pounds per inch

In this module, the forces (elements Force, Friction or Resultant) are added to the objects (elements Block, Pulley, Spring and Point) placing the force in one of the cells around the object. Each object (if there isn't in the chalkboard's border) have 8 cells around it.

This module have 30 elements, presented below. With a description of each one, its data and its number of equations. Notice that some elements have one or two equations, depending of the applied forces. If all forces are horizontal, or vertical, the object have one equation, otherwise have two.

6.1 Stationary reference system

Stationary reference system, with X axis horizontal and positive to the right, and Y axis vertical an positive to upwards.

Equations: None.

Data:

g: Value of the gravity (absolute value). To default FisicaLab write the gravity value to the selected system (9.81 m/s2 to SI, and 32.2 ft/s2 to English system).

6.2 Block

A block, allow one friction force, vertical or horizontal. And don't allow Resultants.

Equations: 1 or 2

Data:

m: Mass block.

6.3 Block above an inclined plane to the left

A block, above an inclined plane to the left in the specified angle. Allow one friction force, parallel to the plane. And don't allow Resultants.

Equations: 2

Data:

m: Mass block.

ang: Angle of the plane, measured from the horizontal.

6.4 Block above an inclined plane to the right

A block, above an inclined plane to the right in the specified angle. Allow one friction force, parallel to the plane. And don't allow Resultants.

Equations: 2

Data:

m: Mass block.

ang: Angle of the plane, measured from the horizontal.

6.5 Springs

Springs without mass, that satisfies the Hooke's law. d is the length that the spring is stretched or compressed. If the spring is stretched d is positive, if is compressed is negative. Allow one or two forces. If have two forces these must have different sense (stretching or compressing the spring), and the same values or unknown data. Don't allow frictions or resultants.

Equations: 1

Data:

k: Spring constant.

d: Length that the spring is stretched or compressed.

6.6 Pulley

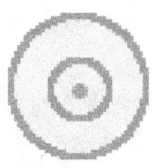

A pulley without mass. Don't allow frictions or resultants.

Equations: 1 or 2

Data:

Name: Name of the pulley (this data is irrelevant).

6.7 Static point

A static point without mass. Don't allow frictions.

Equations: 1 or 2

Data:
Name: Name of the point (this data is irrelevant).

6.8 Angles

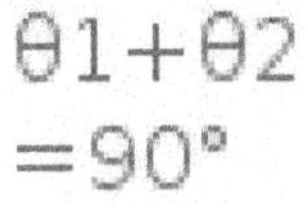

θ1+θ2
=90°

To relate two angles that must be complementary. Both angles must be unknowns.

Equations: 1

Data:
ang1:   One angle.

ang2:   Other angle.

6.9 Forces

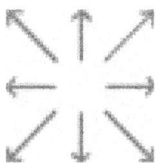

Forces with the direction and sense of the arrow.

Equations: None.

Data:

f   Magnitude of the force.

ang   Positive angle of the force, measured from
                   the horizontal. This entry is displayed only
                   for oblique forces.

6.10 Frictions

Frictions with the direction and sense of the arrow. The oblique frictions,
to be applied to a block on an inclined plane, are assumed parallels to the
plane.

Equations: None.

Data:

Normal: Normal to calculate the friction force.

u: Friction coefficient, static or dynamic.

6.11 Vertical/Horizontal resultant

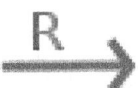

Resultants with the direction and sense of the arrow.

Equations: None.

Data:
f: Magnitude of the resultant.

6.12 Resultant

Oblique resultant.

Equations: None.

Data:
f: Magnitude of the resultant.

ang: Angle of the resultant, measured from the positive X axis. The positive sense is the opposite of clockwise.

7 Examples statics

7.1 Example 1

Calculate the resultant of the three following forces:

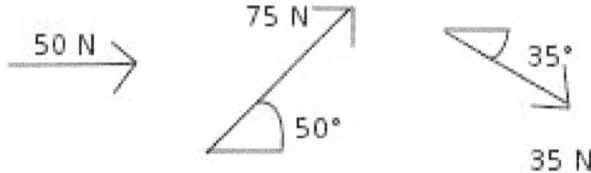

Solution with FisicaLab

Select the Static group and, inside this, the Particles module. Erase the content of the chalkboard and select the SI system. And add one element Point. Now, add two elements Oblique force, one element Horizontal force and one element Resultant. As show the image below:

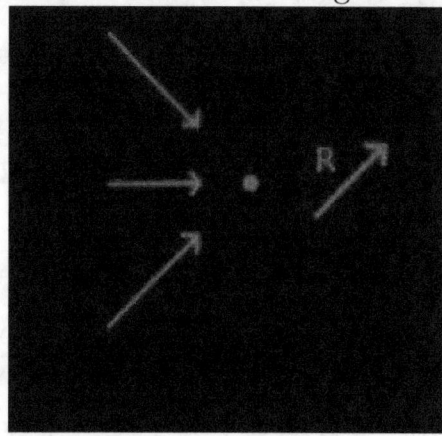

To the element Point we have:

Name   Point

Remember that this name is irrelevant. To the element Horizontal force write:

f   50

To the element Oblique force that pointing upwards, that correspond to the force of 75 Newtons at 50 degrees, write:

f   75

ang   50

To the element Oblique force that pointing downwards, that correspond to the force of 35 Newtons at 35 degrees, write:

f   35

ang   35

And to the element Resultant, write:

f   r

ang   ang

Now click in the icon Solve to get the answer.

```
r = 132.271 N ;   ang = 16.415 degrees ;
Status = success.
```

7.2 Example 2

A block with a mass of 5 kg, is above a 30 degrees inclined plane. A string keep the block in rest (see image). If there isn't friction, which is the value of the stress in the string?.

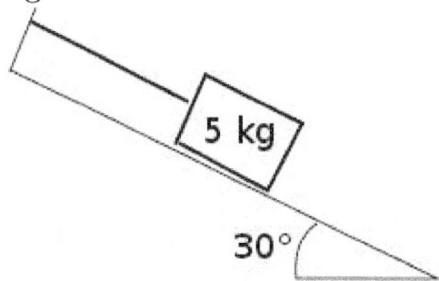

Solution with FisicaLab

Select the Static group and, inside this, the Particles module. Erase the content of the chalkboard and select the SI system. And add two elements Oblique force, one element Block above an inclined plane to the right and one element Stationary reference system, to make the free body diagram of the block. As show the image below:

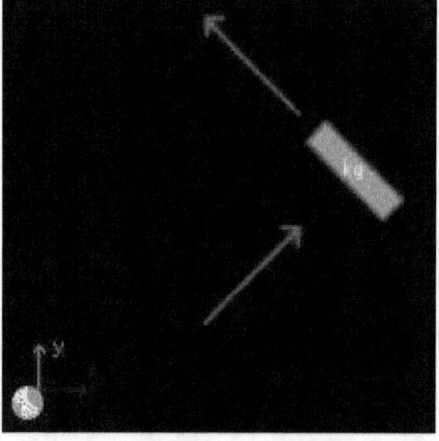

To the Stationary reference system:

g   9.81

To the element Block above an inclined plane to the right write:

m   5
ang   30

To the element Oblique force, that correspond to the normal force, its angle is 60 degrees (if the plane is inclined 30 degrees, the normal force is inclined 60 degrees). Then write:

f   n

ang   60

To the Oblique force that correspond to the stress in the string, the angle is the 30 degrees (this force is parallel to the plane). Then write:

f   t

ang   30

Now click in the icon Solve to get the answer.

```
t = 24.525 N ;   n = 42.479 ;
Status = success.
```

7.3 Example 3

A block, with a mass of 3 kg, is maintained against the wall, to a horizontal force F. If the friction coefficient between the block and wall is 0.25, which is the minimal force F, to avoid the fallen of the block?

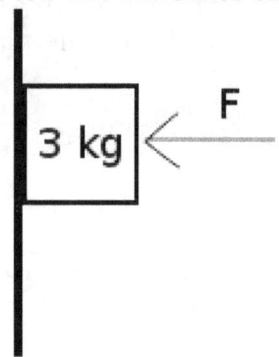

Solution with FisicaLab

The minimal force F, correspond to the case when the friction force is maximum. With this in mind, erase the content of the chalkboard ans select the SI system, and add two elements Horizontal force, one element Friction, one element Block and one element Stationary reference system, to make the free body diagram of the block. As show the image below:

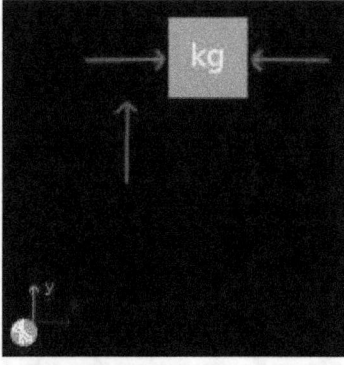

To the Stationary reference system:

g   9.81

To the element Block, write:

m   3

To the element Horizontal force that correspond to the force F, write:

f   F

Because is an unknown data. To the element Horizontal force that correspond to the normal force, write:

f   n

Is an unknown data too. Now, to the element Friction write:

Normal   n

u   0.25

Now click in the icon Solve to get the answer.

```
F = 117.720 N ;   n = 117.720 N ;
Status = success.
```

7.4 Example 4

At what angle will start to slide a block of 3.7 kg when the coefficient of friction is 0.39?

Solution with FisicaLab

Erase the content of the chalkboard and select the SI system, and add one element Stationary reference system, one element Block above an inclined plane (to right or left), one element Force, one element Friction and one element Angles. And make the free body diagram as show the image below:

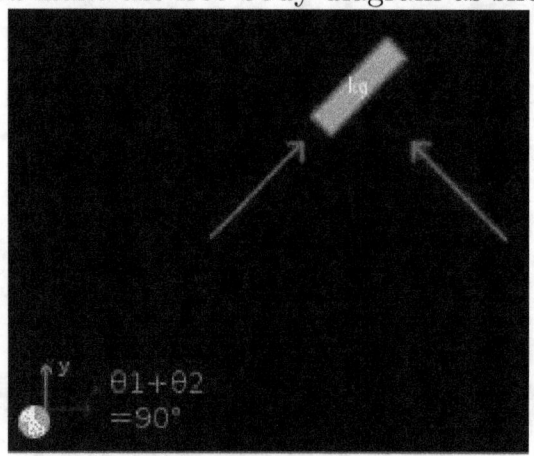

The element Stationary reference system write to default the gravity value. To the element Block above an inclined plane write the mass data. And the angle as an unknown:

m   3.7

ang   ang1

To the element Force, which represent the normal force, both entries are unknown:

f   normal

ang   ang2

And to the element Friction write the normal force and the given coefficient:

Normal   normal

u   0.39

And to the element Angles, add the two unknown angles that must be complementary:

ang1   ang1

ang2   ang2

Now click in the icon Solve to get the answer:

```
ang1 = 21.306 degrees ;   ang2 = 68.694 ;
normal = 33.816 N ;
Status = success.
```

7.5 Example 5

In the system shown in the image, if the stress in the string A is 2 Newtons. Which is the stress in the string B? Which is the mass block?

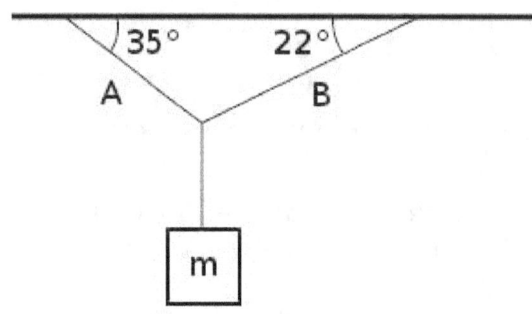

Solution with FisicaLab

Borramos el contenido de la Pizarra, seleccionamos el sistema de unidades SI, y agregamos un elemento Sistema de referencia fijo, un elemento Punto, y un elemento Bloque. Al elemento Punto le agregamos tres elementos Fuerza, y al elemento Bloque solamente uno, para formar dos diagramas de cuerpo libre, tal y como se muestra en la siguiente imagen:

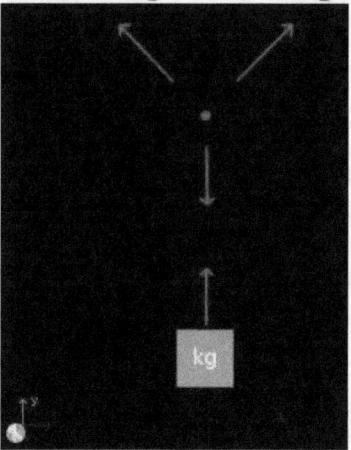

The element Stationary reference system write to default the gravity value. To the element Point the name is irrelevant. To the oblique force, that correspond to the stress in the string A, write:

f   2

ang   35

To the oblique force that correspond to the stress in the string B, write:

f   B

ang   22

Because the magnitude of the force is an unknown data. To the vertical force, applied to the Point, that correspond to the stress in the string that support the block, write:

f   t

Is an unknown data. To the element Block, write:

m   m

Because its mass is an unknown data. To the vertical force, applied to the block, write the same unknown data assigned to the vertical force applied to the Point (is the same string):

f   t

Now click in the icon Solve to get the answer.

```
t = 1.809 N ;   B = 1.767 N ;   m = 0.184 kg ;
Status = success.
```

7.6 Example 6

A block of 5 kg is supported to a spring as show the image. If the spring constant is 500 N/m. Which is the stress in the string? How much the spring is stretched?

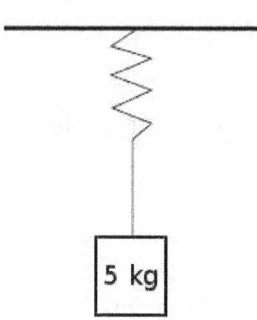

Solution with FisicaLab

Erase the content of the chalkboard and select the SI system, and add one element Spring, one element Block, and one element Stationary reference system. Then add one element Force to the element Spring, and one element Force to the element Block, to make two free body diagrams as show the image below:

The element Stationary reference system write to default the gravity value. To the element Block, write:

m   5

To the vertical force applied to the spring, that is an unknown data, write:

f   t

To the element Spring, the length stretched is an unknown data, write:

k   500
d   d

   To the vertical force applied to the block, that is an unknown data and
the same unknown data that to the other force, write:

f   t

   Now click in the icon Solve to get the answer.

```
t = 49.050 N ;   d = 0.098 m ;
Status = success.
```

7.7 Example 7

A spring support a block of unknown mass, see image below. If the constant spring is 150 N/m, and this is stretched 5 cm, Which is the mass of the block in grams?

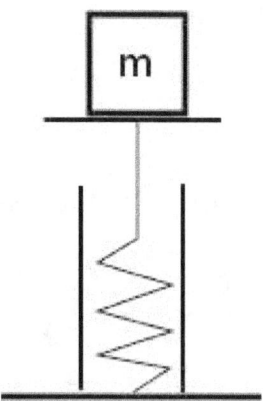

Solution with FisicaLab

Erase the content of the chalkboard and select the SI system, and add one element Spring, one element Block, and one element Stationary reference system. Then add one element Force to the element Spring, and one element Force to the element Block, to make two free body diagrams as show the image below:

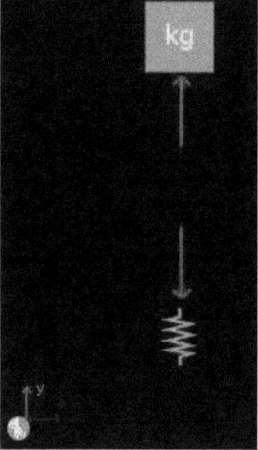

The element Stationary reference system write to default the gravity value. To the element Block, write:

m   m @ g

Because the mass is an unknown data, and we want this in grams. To the vertical force applied to the block, that is an unknown data, write:

f   f

To the element Spring, write:

k   150

d   -5 @ cm

The negative sign is because the spring is stretched. To the vertical force applied to the block, that is an unknown data and the same unknown data that to the other force, write:

f   f

Now click in the icon Solve to get the answer.

```
f = 7.500 N ;   m = 764.526 g ;
Status = success.
```

7.8 Example 8

In the system shown in the image. Which is the value of the force F?

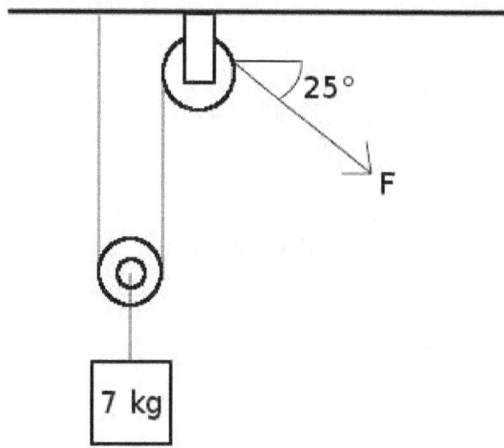

Solution with FisicaLab

Erase the content of the chalkboard and select the SI system, and add the necessary elements to make the three free body diagrams as show the image below (The white circles are just for clarify):

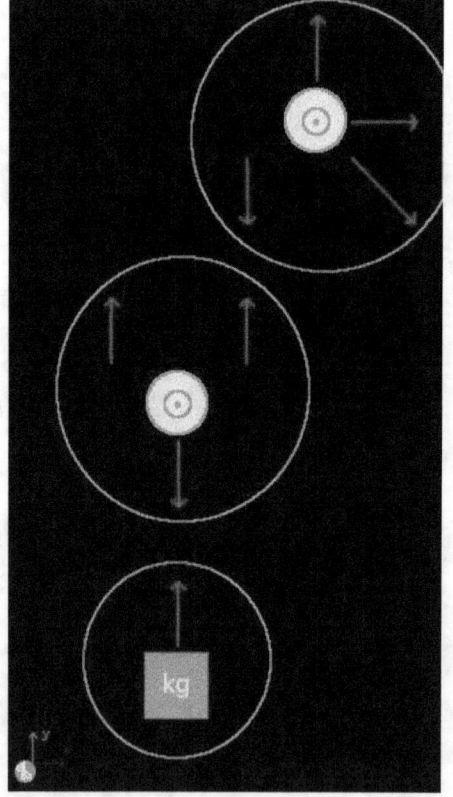

To the element Block, write:

m   7

To the element force applied to the block, its values is an unknown data, calling this t1, write:

f   t1

To the elements Pulley the names are irrelevant. To the downwards vertical force applied to the pulley immediately above the block, write:

f   t1

Is the same string that support the block. In that pulley, the two upwards vertical forces, are unknown data, and the same unknown data, because is the same string. Calling this t2, write to both forces:

f   t2

In the other pulley, the downwards vertical force, and the oblique force are the same string, then to the vertical force write:

f   t2

And to the oblique force, write:

f   t2
ang   25

The others two forces are the reactions that support the pulley. Calling Ry the vertical reaction, write:

f   Ry

And calling Rx the horizontal reaction, write:

f   Rx

Now click in the icon Solve to get the answer.

    t2 = 34.335 N ;   Ry = 48.846 N ;   t1 = 68.670 N ;

```
Rx = -31.118 N ;
Status = success.
```

The negative sign in the Rx data is because the sense of this force, is the opposite of the force that we add to the pulley.

7.9 Example 9

A block with a mass of 20 kg, is supported to a string, as shown the image. If the spring is stretched 4 centimeters, which is the constant spring? which are the reactions in A?

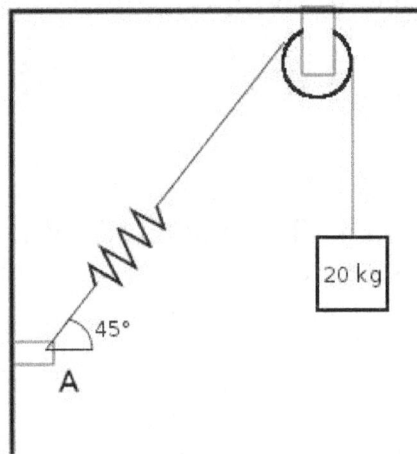

Solution with FisicaLab

Erase the chalkboard and add the necessary elements to make three free body diagrams, as show the image below:

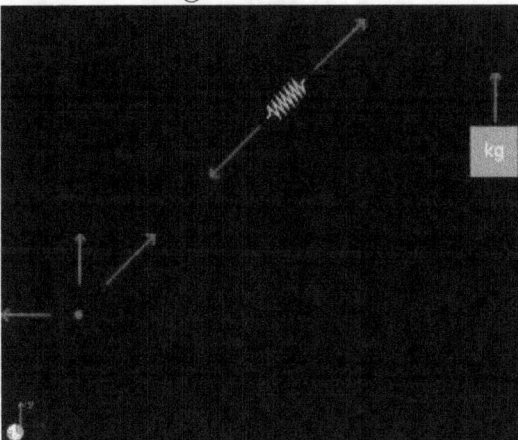

We don't need the reactions on the pulley, for that we don't add this. To the element Block, write:

m   20

To the force applied to the block:

f   t

To the spring write:

k   k

d   4 @ cm

And to the forces applied to the spring, write in both:

f   t

ang   45

Here the angle is irrelevant. The magnitude of the oblique force applied to the point, is an unknown data, but is the same unknown data applied to the spring. Then write:

f   t

ang   45

We calling the reactions forces applied to the point, Ry and Rx, respectively. Then write:

f   Ry

f   Rx

Now click in the icon Solve to get the answer.

```
t = 196.200 N ;   k = 4905.000 N/m ;   Rx = 138.734 N ;
Ry = -138.734 N ;
Status = success.
```

The negative sign in the Ry data is because the sense of this force, is the opposite of the force that we add to the pulley.

8 Module rigid statics

The conversion factors for SI system are:

cm	centimeter
mm	millimeter
ft	feet
in	inch
g	gram
slug	slug
T	metric ton
N*cm	Newton centimeter

And for the English system are:

kg	kilogram
g	gram
m	meter
cm	centimeter
mm	millimeter
in	inch
lb*in	pounds inch

In this module, the forces (elements Force, Friction and Resultant) are applied to elements object (elements Point, elements of beam or elements of solid), placing these at adjacent cells to the objects. Each element object, if isn't placed at the chalkboard's border, has 8 adjacent cells. This module have 27 elements, presented below. With a description of each one, its data and its number of equations. Notice that some elements add two or three equations, depending on the forces applied to it. If all forces are horizontals, or verticals, two equations are added, otherwise three equations are added.

8.1 Stationary reference system

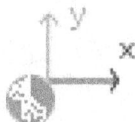

Stationary reference system, with X axis horizontal and positive to the right, and Y axis vertical an positive to upwards.

Equations: None.

Data:

g: Value of the gravity (absolute value). To default FisicaLab write the gravity value to the selected system (9.81 m/s2 to SI, and 32.2 ft/s2 to English system).

8.2 Points

Set a rigid body, without mass, composed by elements Point. Not support forces, frictions or resultants.

Equations: 2 or 3

Data:

Name: Name of the body.

8.3 Beam

Set a rigid body (a beam) composed by elements of beam. Not support forces, frictions or resultants. Starting from a horizontal beam, the left side is the vertex to measure the angle.

Equations: 2 or 3. If all applied forces are vertical, are two equations (the sum of vertical forces and sum of moments). Isn't allowed apply horizontal forces only, because FiscaLab assumes that the beam have a weight.

Data:

Name: Name of the beam.

m: Mass of the beam.

lc: Distance from the center of mass of the beam to the vertex point (the distance is measured along the beam).

ang: Angle of inclination of the beam, measured from the positive X axis. The positive sense is the opposite of clockwise.

8.4 Solid

Set a rigid body (a solid of any shape) composed by elements of solid. Not support forces, frictions or resultants. If the solid is placed above an inclined surface, the coordinates X and Y of the mass center, and the coordinates X and Y of all it's elements, are measured from an X axis, parallel at surface and positive to upwards, and a Y axis perpendicular to surface and positive to upwards.

Equations: 2 or 3. If all applied forces are vertical, are two equations (the sum of vertical forces and sum of moments). Isn't allowed apply horizontal forces only, because FiscaLab assumes that the solid have a weight.

Data:

Name: Name of the solid.

m: Mass of the solid.

xc: X coordinate of the center of mass.

yc: Y coordinate of the center of mass.

ang: Angle of the surface where the solid is placed. Is measured from the positive X axis and the positive sense is the opposite of clockwise.

8.5 Point

Element to a body without mass (Points), can be added any number of points to make the body. Support forces, frictions or resultants.

Equations: None.

Data:
Points: Name of the rigid body to which it belongs.

x: X coordinate of the point.

y: Y coordinate of the point.

8.6 Angles

$\theta 1 + \theta 2$
$= 90°$

To relate two angles that must be complementary. Both angles must be unknowns.

Equations: 1

Data:
ang1:   One angle.

ang2:   Other angle.

8.7 Elements of beam

Element to a beam, can be added any number of elements to make the beam. All have the same properties, only change the image for visual purposes (know when a beam is vertical, horizontal or have a slope). Support forces, frictions or resultants.

Equations: None.

Data:

Beam: Name of the beam to which it belongs.

l: Distance from the element to the vertex point.

8.8 Elements of solid

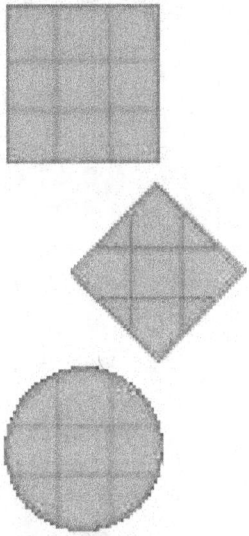

Element to a solid, can be added any number of elements to make the solid. All have the same properties, only change the image for visual purposes (know when a part is on the edge or inside the solid, or if is a curved or inclined surface). Support forces, frictions or resultants.

Equations: None.

Data:

Solid: Name of the solid to which it belongs.

x: X coordinate of the element.

y: Y coordinate of the element.

8.9 Forces

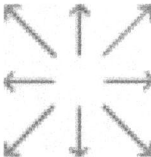

Forces with the direction and sense of the arrow.

Equations: None.

Data:

f   Magnitude of the force.

ang   Positive angle of the force, measured from the horizontal. This entry is displayed only for oblique forces.

8.10 Frictions

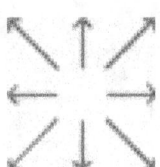

Frictions with the direction and sense of the arrow.

Equations: None.

Data:

Normal: Normal to calculate the friction force.

ang: Positive angle of the force, measured from the horizontal. This entry is displayed only for oblique forces.

u: Friction coefficient, static or dynamic.

8.11 Couple

Moment of a couple of forces.

Equations: None.

Data:

m: Moment of the couple. A positive value corresponds to a direction opposite to clockwise.

8.12 Beams of 2 forces

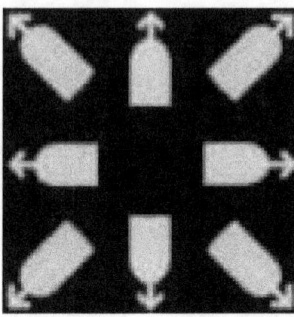

Force applied by a beam of 2 forces. This elements have similar characteristics of forces. The only difference is that if the data of the force is unknown, the result will display if the beam is in tension or compression. Since this element represent the force applied by the beam, a positive value means compression and a negative means tension.

Equations: None.

Data:

f: Magnitude of the force. If this data is unknown, FisicaLab will display in the result if the beam is in tension or compressed.

ang: Positive angle of the force, measured from the horizontal. This entry is present only in oblique beams.

8.13 Truss

Simple truss.

Equations: None.

Data:
Name: Name of the truss.

8.14 Joint

Joint of truss. Support forces and beams of truss.

Equations: 2

Data:
Truss: Name of the truss that owns the joint.

8.15 Beams of truss

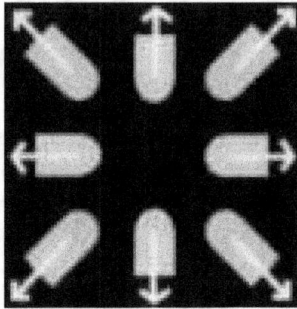

Beams of truss.

Equations: None.

Data:

t: Magnitude of the force inside the beam. A positive value means tension and a negative value means compression. If this data is unknown, FisicaLab will display in the result if the beam is in tension or compressed.

ang: Positive angle of the force, measured from the horizontal. This entry is present only in oblique beams.

8.16 Resultant

General resultant.

Equations: None.

Data:

mr: Resultant moment. A positive value corre-
               sponds to a sense opposite to clockwise.

fr: Magnitude of the resultant force.

ang: Angle of the resultant force, measured from
               the positive X axis. The positive sense is
               the opposite of clockwise.

8.17 Resultant with horizontal force

Resultant with horizontal force (positive to the right and negative to the
left).

Equations: None.

Data:

mr: Resultant moment. A positive value corre-
               sponds to a sense opposite to clockwise.

fr: Magnitude of the resultant force.

8.18 Resultant with vertical force

Resultant with vertical force (positive to upwards and negative to downwards).

Equations: None.

Data:

mr: Resultant moment. A positive value corresponds to a sense opposite to clockwise.

fr: Magnitude of the resultant force.

9 Examples statics rigid bodies

9.1 Example 1

Calculate the resultant, at point A, of the forces applied to the piece.

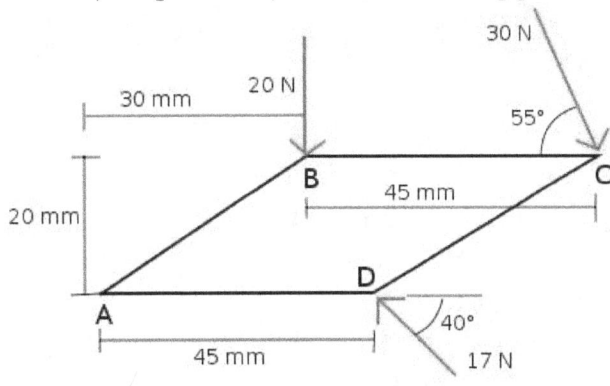

Solution with FisicaLab

Select the group Static and, inside this, the module Static of solids. Erase the content of the chalkboard and select the SI system. And add one element Points, four elements Point, one element Resultant and three elements Force, as show the image below:

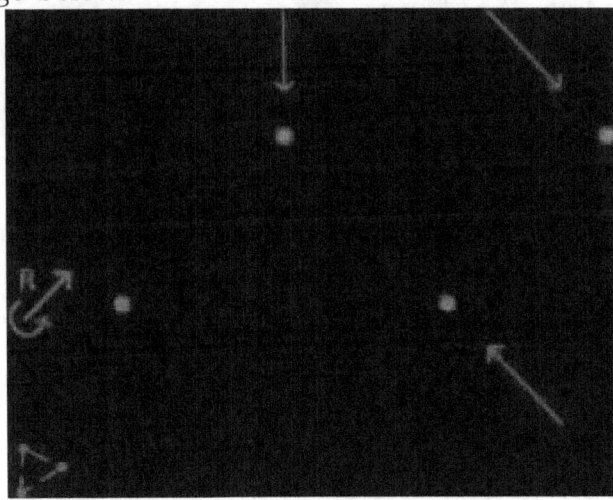

To the element Points, we call this *Piece*:

Name   Piece

To the element Point that represent the corner A, assuming this is at the origin, we have:

Name   Piece

x   0

y   0

Now to the element Point at corner B:

Name	Piece
x	30 @ cm
y	20 @ cm

And to the applied force at this corner:

f   20

To the element Point at corner C:

Name	Piece
x	75 @ cm
y	20 @ cm

And to the applied force here:

f   30

ang   55

To corner D:

Name	Piece
x	45 @ mm
y	0

And to the force:

f   17

ang   40

Now to the element Resultant:

M   Mr

f   Fr

ang   angr

Now click in the icon Solve to get the answer:

```
Mr = -2.296 N*m ;  Fr = 33.906 N ;  angr = 277.089 degrees ;
Status = success.
```

9.2 Example 2

Calculate the resultant, at point O, of the applied force (ignore the weight of the beam).

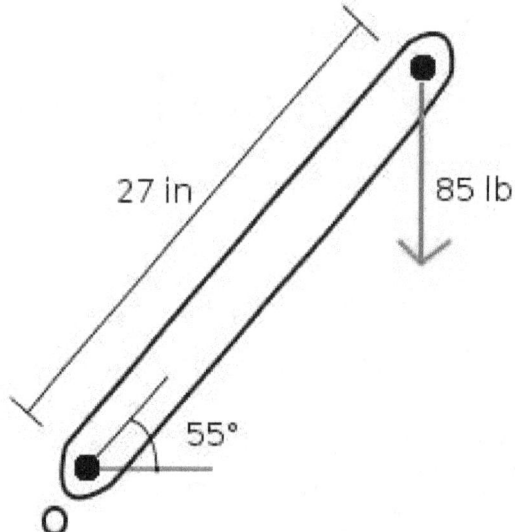

Solution with FisicaLab

Select the group Static and, inside this, the module Static of solids. Erase the content of the chalkboard and select the English system. And add one element Stationary reference system, one element Beam, two elements of beam, one element Force and one element Resultant with vertical force, as show the image below:

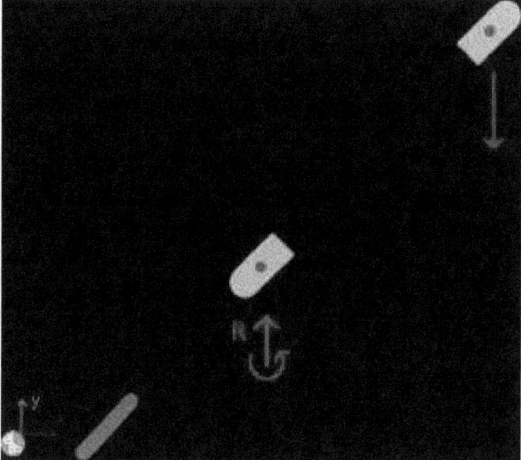

The element Stationary reference system set, to default, the gravity value:

g   32.2

To the element Beam, we call this *Beam*, we have:

Name	Beam
m	0
lc	0
ang	55

Here we have 0 to the mass, because we must ignore the weight of the beam. To the element of beam that represent the lower side, we have:

Name	Beam
l	0

And to the element Resultant applied here:

M	Mr
f	Fr

To the other element of beam, we have:

Name	Beam
l	27 @ in

And to the applied force:

f	85

Now click in the icon Solve to get the answer:

```
Mr = -109.696 lb*ft ;   Fr = -85.000 lb ;
Status = success.
```

9.3 Example 3

If we want to replace the two forces of 175 N and 380 N force with a single force applied at point P as shown the image, What should be the distance from point P to point O and the magnitude of this force? (The beam and the drum form a rigid body)

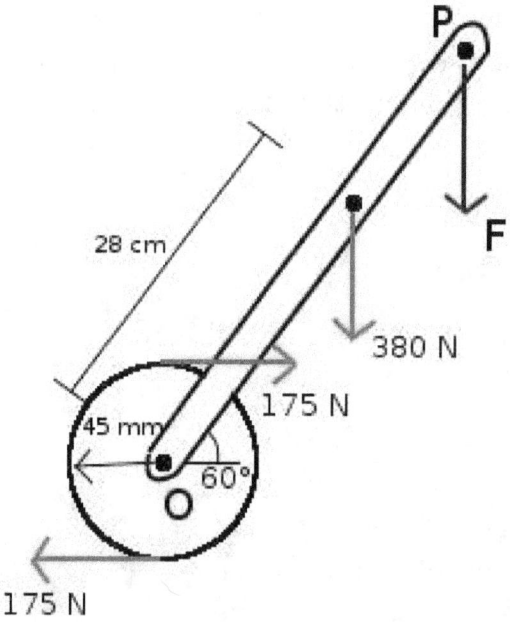

Solution with FisicaLab

Select the group Static and, inside this, the module Static of solids. Erase the content of the chalkboard and select the SI system. This problem can be solved on different ways, here we will solve it as follow. First add one element Points, three elements point and the necessary elements force and resultant, as show the image below. Our initial intention is calculate the couple of the pair of 175 N forces, at point O.

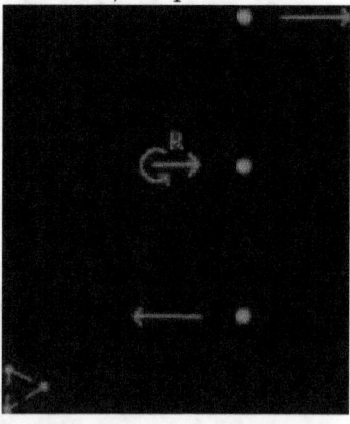

To the element Points, that represent the different points of the drum and that we call *Drum*, we have:

Name   Drum

To the element Point that represent the center O, we assume this at origin:

Name   Drum

x   0

y   0

And to the element Resultant applied here:

M   Mr

f   Fr

Now to the upper point, we have:

Name   Drum

x   0

y   45 @ mm

And to the lower point:

Name   Drum

x   0

y   -45 @ mm

And to the two applied forces:

f   175

Now click in the icon Solve to get the answer:

```
Mr = -15.750 N*m ;  Fr = 0.000 N ;
Status = success.
```

Now that we know the moment of the pair of 175 N forces, erase the content of chalkboard and add the necessary elements as show the image below:

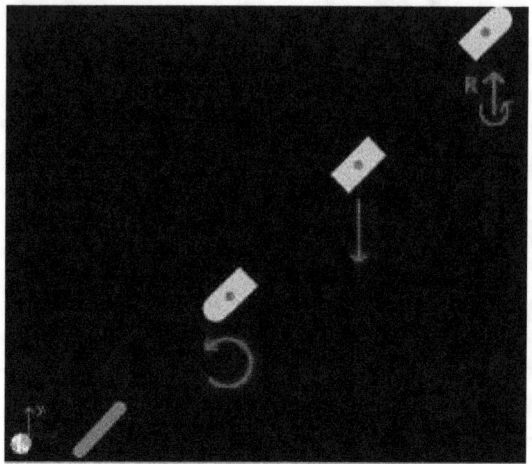

The Stationary reference system set the gravity value to default. Now to the element Beam, we call this *Beam*, we have:

Name	Beam
m	0
lc	0
ang	60

Here we ignore the weight of the beam. To the element of beam that represent the lower side and to the applied couple (the previous calculated value), we have:

Name	Beam
l	0
M	-15.750

Now to the element of beam where is applied the other force, and to this force, we have:

Name	Beam
l	28 @ cm
f	380

To the element of beam that represent the point P:

Name   Beam
l   d

Here the distance is an unknown data. And to the resultant applied here:

M   0
f   Fr

Here the couple is 0, because we want just a force, which is an unknown data. Now click in the icon Solve to get the answer:

```
d = 0.363 m ;  Fr = -380.000 N ;
Status = success.
```

9.4 Example 4

Calculate the resultant, at joint O, of the applied forces.

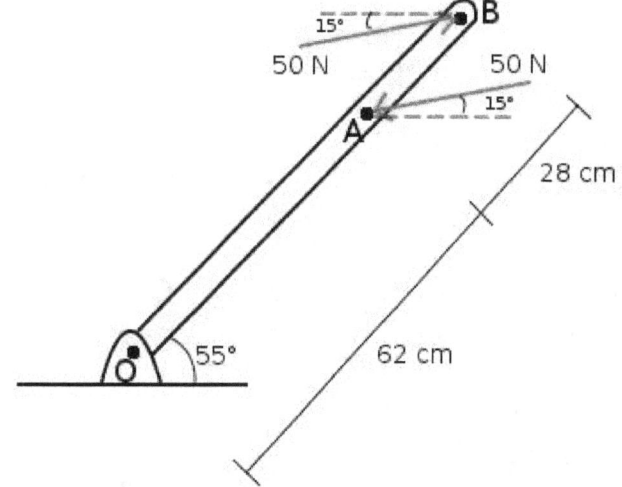

Solution with FisicaLab

Select the group Static and, inside this, the module Static of solids. Erase the content of the chalkboard and select the SI system. And add one element Stationary reference system, one element Beam, three elements of beam and the necessary elements Force and Resultant, as show the image below:

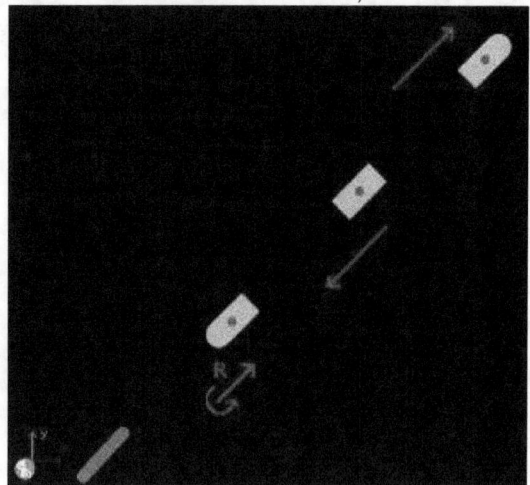

The gravity value is added to default. Now to the element Beam, that we call *Beam*, we have:

Name	Beam
m	0
lc	0
ang	55

To the element of beam that represent the lower side, and to the resultant applied here, we have:

Name	Beam
l	0

M	Mr
f	Fr
ang	angr

On the resultant element all are unknown data. Now to the element of beam that represent the point A, and to the applied force, we have:

Name	Name
l	62 @ cm

f	50
ang	15

To the element of beam that represent the point B and to the applied force, we have:

Name	Beam
l	90 @ cm

f	50
ang	15

Now click in the icon Solve to get the answer:

```
Mr = -8.999 N*m ;   Fr = 0.000 N ;   angr = 208.262 degrees ;
Status = success.
```

As we expected the resultant is only a couple. Here the angle is irrelevant.

9.5 Example 5

Calculate the support reactions, knowing that the beam has a mass of 4 kg and a length of 2.1 m.

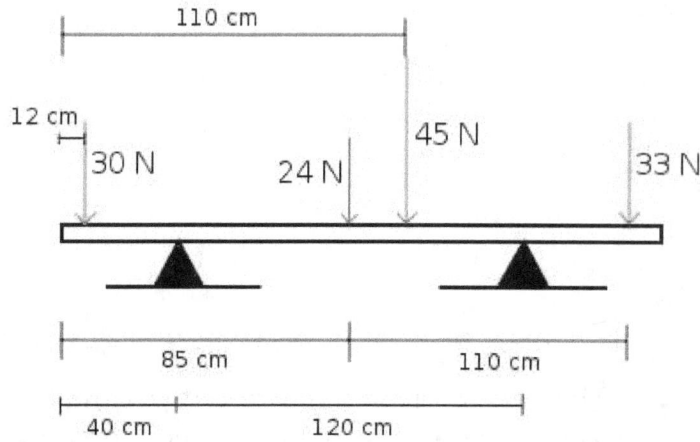

Solution with FisicaLab

Select the group Static and, inside this, the module Static of solids. Erase the content of the chalkboard and select the SI system. And add one element Stationary reference system, one element Beam, six elements of beam and the necessary elements Force, as show the image below:

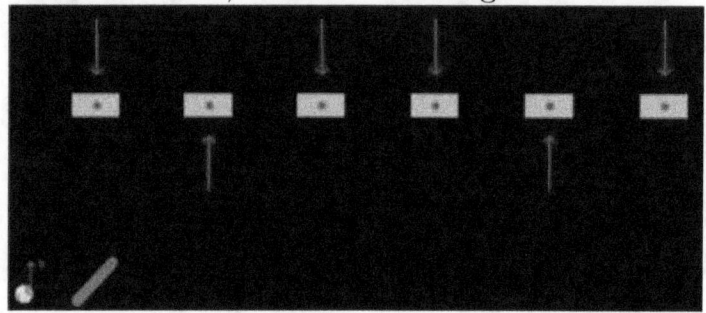

The Stationary reference system set the gravity value to default. Now to the element Beam, that we call *Beam* and assuming has its center of mass at middle, we have:

Name	Beam
m	4
lc	1.05
ang	0

Now to the elements of beam and its applied forces we have, from left to right:

Name	Beam

l	12 @ cm

f	30

To the element at 40 cm from the left side, we call RA the reaction:

Name	Beam
l	40 @ cm

f	RA

At 85 cm from the left side:

Name	Beam
l	85 @ cm

f	24

At 110 cm:

Name	Beam
l	110 @ cm

f	45

At 160 cm, we call RB the reaction:

Name	Beam
l	160 @ cm

f	RB

And at 195 cm, the last element:

Name	Beam

l   195 @ cm

f   33

Now click in the icon Solve to get the answer:

```
RB = 92.130 N ;   RA = 79.110 N ;
Status = success.
```

9.6 Example 6

If the tension in the cable BC is 155 N, what are the reactions at A? (The mass of the beam is 3 kg and its center of mass is 75 cm from A).

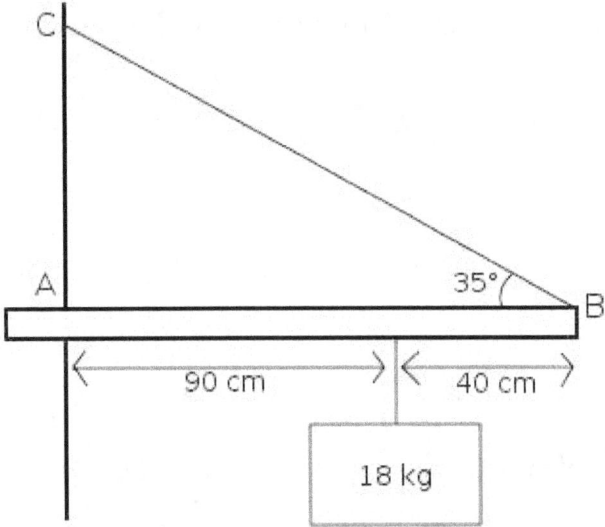

Solution with FisicaLab

Select the group Static and, inside this, the module Static of solids. Erase the content of the chalkboard and select the SI system. And add one element Stationary reference system, one element Beam, three elements of beam and the necessary elements Force and Couple, as show the image below:

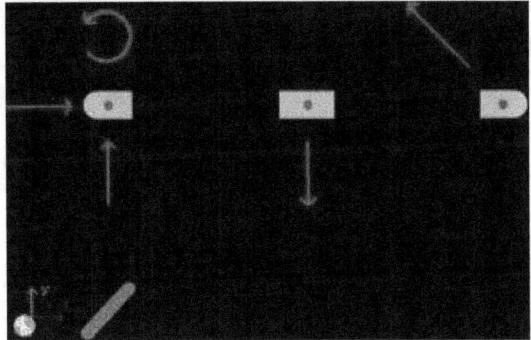

The Stationary reference system set the gravity value to default. Now to the element Beam, that we call *Beam*:

Name	Beam
m	3
lc	75 @ cm
ang	0

To the element of beam at side A, we have:

Name   Beam

l   0

And to the forces and couple applied, all unknown data, we have:

f   Ry

f   Rx

M   M

To the element of beam where the mass is hanging:

Name   Beam

l   0.9

And to the applied force, the weight of the mass (18 kg*9.81 m/s2 = 176.58 N), we have:

f   176.58

And to the element of beam that represent the right side:

Name   Beam

l   1.3

And to the applied force by the cable:

f   155

ang   35

Now click in the icon Solve to get the answer:

```
Ry = 117.106 N ;   Rx = 126.969 N ;   M = 65.419 N*m ;
Status = success.
```

9.7 Example 7

What is the maximum force F that can be applied without tipping the piece? (The mass of the piece is 3.5 kg and the mass center point is indicated by the 18 cm).

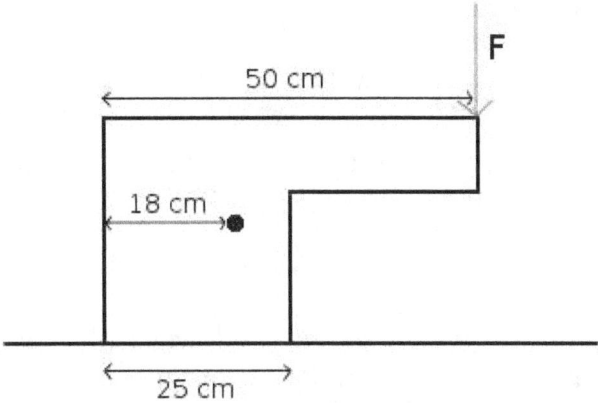

Solution with FisicaLab

Select the group Static and, inside this, the module Static of solids. Erase the content of the chalkboard and select the SI system. And add one element Stationary reference system, one element Solid, two elements of solid and the necessary elements Force, as show the image below:

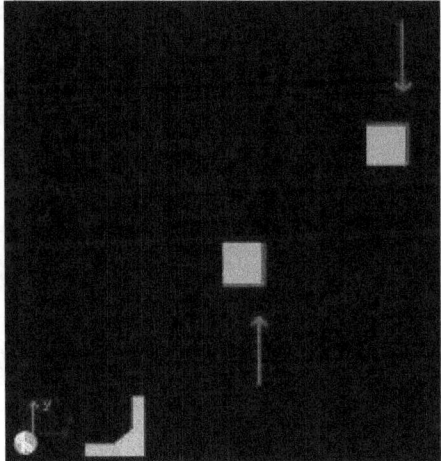

The Stationary reference system set the gravity value to default. Now to the element Solid, that we call *Piece*, we have:

Name	Piece
m	3.5
xc	18 @ cm
yc	0
ang	0

To the element of solid that represent the corner where is applied the force F, we have:

Name	Piece
x	50 @ cm
y	0

And to the applied force here, that is an unknown data:

f	F

The other element of solid represent the lower right corner, because when the tipping is imminent, the normal force is applied here. Therefore, we have:

Name	Piece
x	25 @ cm
y	0

And to the applied force, that represent the normal force and that is an unknown data:

f	N

Now click in the icon Solve to get the answer:

```
N = 43.949 N ;   F = 9.614 N ;
Status = success.
```

9.8 Example 8

A uniform block of 4 kg, as show the image, is placed on a slope. What is the maximum angle of inclination of the plane, so that the block does not slide or tipping (whichever comes first)? (The coefficient of friction between the block and the plane is 0.4).

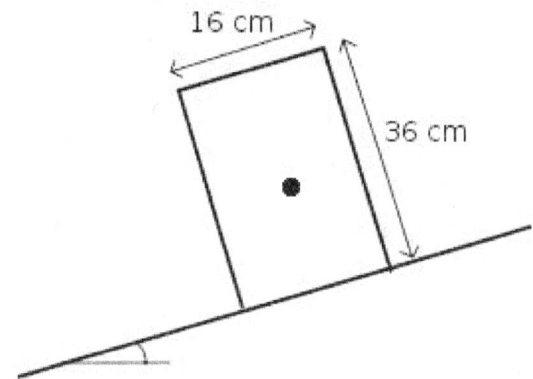

Solution with FisicaLab

Select the group Static and, inside this, the module Static of solids. Erase the content of the chalkboard and select the SI system. And add one element Stationary reference system, one element Solid, one element of Solid, one element Angles and the necessary elements Force and Friction, as show the image below:

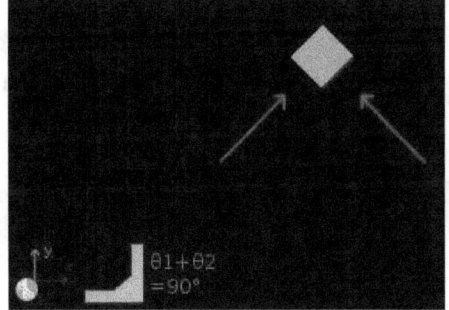

Our intention to introduce a frictional force is find the angle when the slip is imminent. Thus, the block element represents the point where is applied the normal and here we apply the friction force too (we can move the friction along its action line, so we can apply it in this element).

The Stationary reference system set the gravity value to default. Now on the element Solid, that we call *Block*, we have:

Name	Block
m	4
xc	8 @ cm
yc	18 @ cm

ang   ang1

Here the angle is an unknown data. To the element of solid, we have:

Name   Block

x   x @ cm

y   0

The x coordinate is unknown. This must be so. Because when the sliding of block is imminent, we don't know at where is applied the normal force. Now to the normal force:

f   normal

ang   ang2

To the element Friction:

Normal   normal

ang   ang1

u   0.4

The angle is the same as of the block. Because the friction is parallel to the plane. Now to the element Angles, we have:

ang1   ang1

ang2   ang2

Because these angles must be complementary. Once entered the data, click on the icon Solve to get the answer:

```
normal = 36.433 N ;  ang1 = 21.801 degrees ;
ang2 = 68.199 degrees ;  x = 0.800 cm ;
Status = success.
```

Then, the angle when the sliding of block is imminent is 21.801 degrees (As we expected, because `arctan 0.4 = 21.801`). In the case of a negative value, this would mean that the block has already overturned.

9.9 Example 9

A uniform ladder, of 3.5 kg and 3.8 m long, rests against a wall as show the image. If a child with a mass of 40 kg, ups to the fourth step at 3.2 m from the base, find the angle for which this position corresponds to the slip imminent if the coefficient of friction between the ladder and the floor is 0.35 (there is no friction between the ladder and the wall).

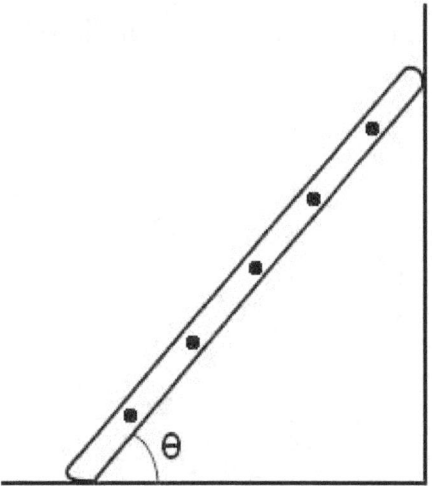

Solution with FisicaLab

Select the group Static and, inside this, the module Static of solids. Erase the content of the chalkboard and select the SI system. And add one element Stationary reference system, one element Beam, three elements of beam and the necessary elements Force and Friction, as show the image below:

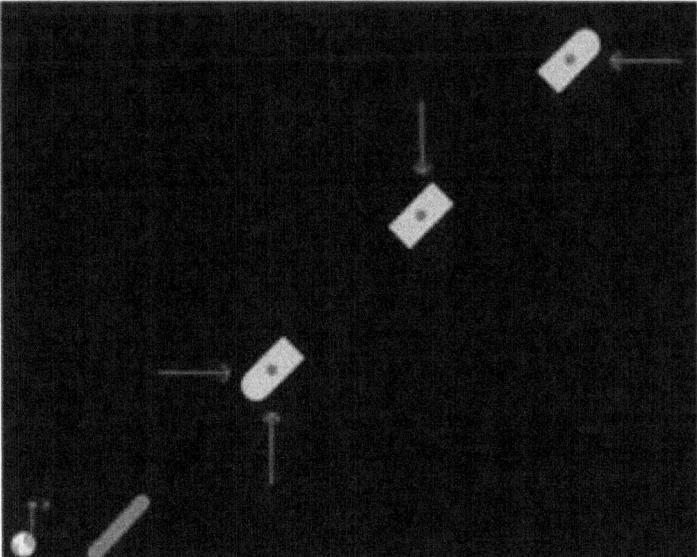

The Stationary reference system set the gravity value to default. Now to the element Beam, that we call *Ladder*, recalling that the ladder is uniform and that the angle is unknown:

Name	Ladder
m	3.5
lc	1.9
ang	ang

To the element of beam that represent the lower side, we have:

Name	Ladder
l	0

And to the normal force and the friction:

f	N

Normal	N
u	0.35

Now to the element that represent the point where the child is supported, we have:

Name	Ladder
l	3.2

And to the force that represent the weight of this child (40 `kg*9.81 m/s2` = `392.4 N`):

f	392.4

Finally, to the element that represent the upper side:

Name	Ladder
l	3.8

And to the force that represent the reaction of the wall:

f	Rx

Now click in the icon Solve to get the answer:

```
Rx = 149.357 N ;  ang = 66.748 degrees ;  N = 426.735 N ;
Status = success.
```

9.10 Example 10

Consider the structure in the image below. Calculate reactions at O and reactions between the beams at points A, B and C. Ignore the weight of the beams.

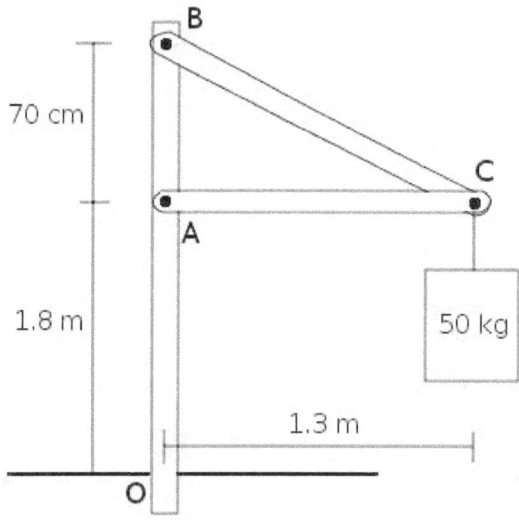

Solution with FisicaLab

Select the group Static and, inside this, the module Static of solids. Erase the content of the chalkboard and select the SI system. And add one element Stationary reference system, three elements Beam, seven elements of beam and the necessary elements Force and Couple, as show the image below (the yellow lines separate the areas where are defined the beams):

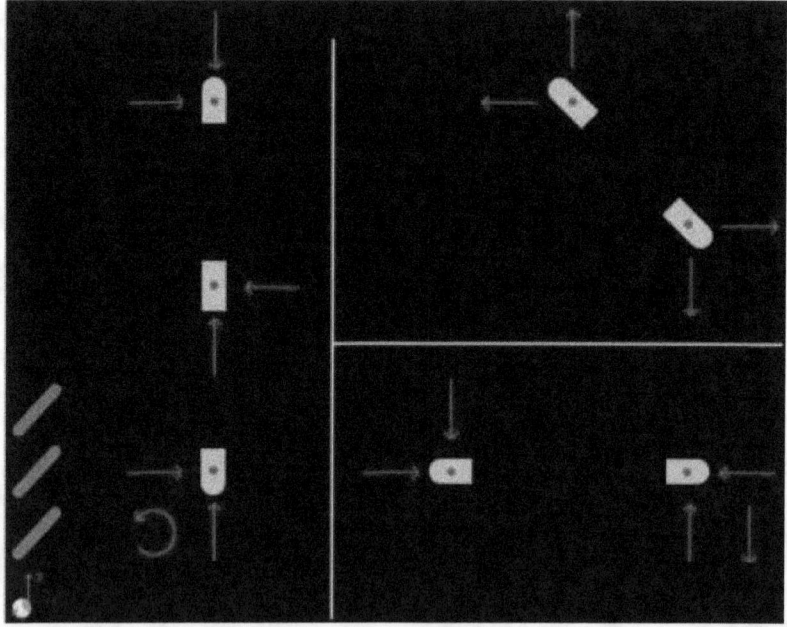

In the image can be seen that, according to Newton's third law, the reactions between the beams have a different sense. And we assign the same magnitude or unknown data. The Stationary reference system set the gravity value to default. Now to the element Beam that represent the vertical beam, that we call *OAB*:

Name	OAB
m	0
lc	0
ang	90

This beam is composed of three elements, which represent the points O, A and B. For the element in O, and their respective reactions, called Rx, Ry, M, we have:

Name	OAB
l	0
f	Rx
f	Ry
M	M

For the element in A and their respective reactions, RAx and RAy, we have:

Name	OAB
l	1.8
f	RAx
f	RAy

For the element in B and their respective reactions, RBx and RBy, we have:

Name	OAB
l	2.5
f	RBx
f	RBy

Now for the element Beam representing the horizontal beam, that we call *AC*, we have:

Name	AC
m	0
lc	0
ang	0

For the element of beam that represent the point A, and to its respective reactions, we have:

Name	AC
l	0
f	RAx
f	RAy

To the element of beam at point C, for the reactions RCx, RCy and to the weight of 50 kg (50 kg*9.81 m/s2 = 490.5 N), we have:

Name	AC
l	1.3
f	RCx

f	RCy

f	490.500

Now to the element Beam that represent the beam from B to C, that we call *BC* (here the angle is entered as slope -0.7/1.3):

Name	BC
m	0
lc	0
ang	-28.301

To the element of beam at B and to its reactions, we have:

Name	BC
l	0

f	RBx

f	RBy

For the element of beam at C, we should calculate the distance between B and C. Which can be calculated with `hypot(0.7,1.3)`:

Name	BC
l	1.476

And to the reactions RCx and RCy (we don't add here the weight of the 50 kg, because was added at the beam AC):

f	RCx

f	RCy

Now click in the icon Solve to get the answer:

```
RAx = 910.919 N ;   RAy = -0.000 N ;   RCy = 490.500 N ;
RCx = 910.919 N ;   RBx = 910.919 N ;   RBy = 490.500 N ;
M = 637.643 N*m ;   Ry = 490.500 N ;   Rx = 0.000 N ;
Status = success.
```

Note: *A more easy way to solve this problem (that avoid having to calculating the length of the beam BC), is replace the BC beam with two elements Beam of 2 forces (notice that the beam BC is an element that supports two forces). See image below. Everything is a matter of deciding which is best for the problem to solve. The beam AC is also a beam of two forces. However, we can't replace both.*

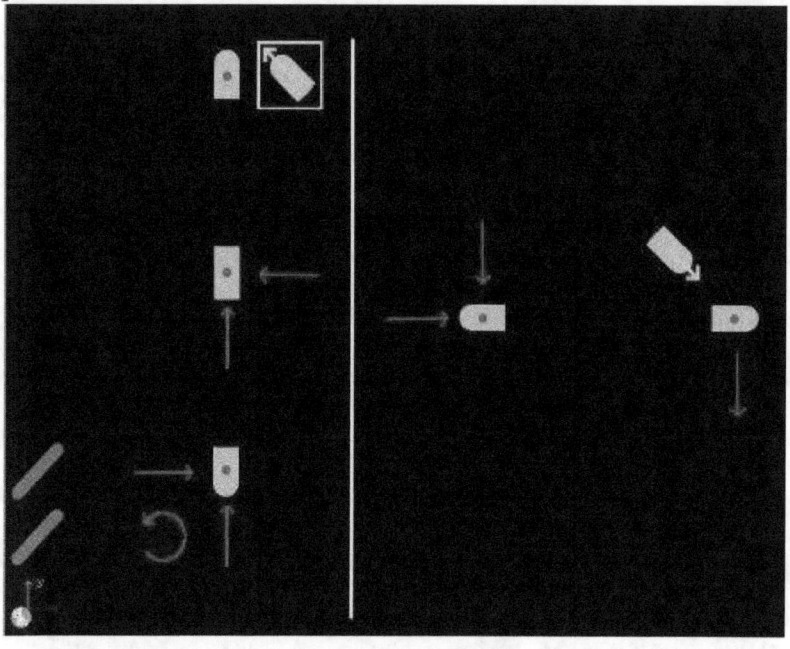

9.11 Example 11

A uniform block of 25 kg and 50 cm of edge, as show the image below, is placed on a slope of 20 degrees. If we want move up the block as fast as possible (constant speed) by the horizontal force F, What is the maximum possible coefficient of friction? Which is the force?

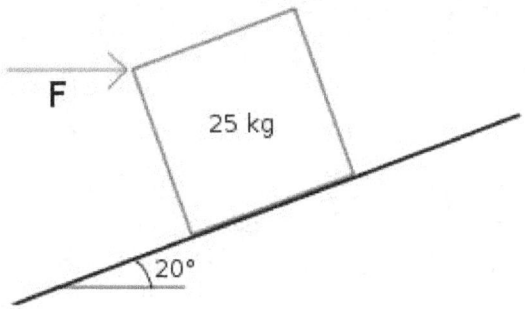

Solution with FisicaLab

Select the group Static and, inside this, the module Static of solids. Erase the content of the chalkboard and select the SI system. And add one element Stationary reference system, one element Solid, two elements of solid and the necessary elements Force and Friction, as show the image below:

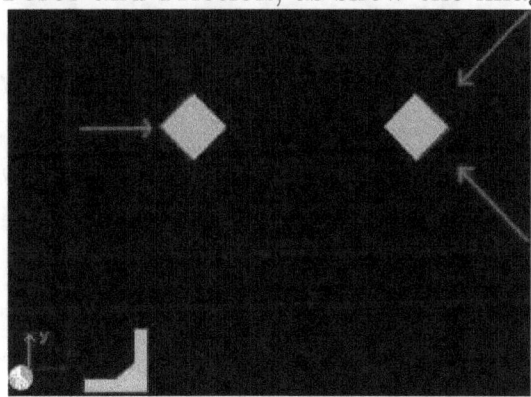

The Stationary reference system set the gravity value to default. Now to the element Solid, that we call *Block* we have:

Name	Block
m	25
xc	25 @ cm
yc	25 @ cm
ang	20

To the element of solid that represent the upper corner, where is applied the force F, we have:

Name	Block
x	0
y	50 @ cm

And to the horizontal force:

f	F

The normal force and the friction are applied to the element of solid that represent the lower right corner. Because when we move up the block as fast as possible, the tipping is imminent. Then, we have:

Name	Block
x	50 @ cm
y	0

To the force that represent the normal force:

f	N
ang	70

And to the friction:

Normal	N
ang	20
u	u

Now click in the icon Solve to get the answer:

```
F = 262.970 N ;  N = 320.401 N ;  u = 0.509 ad ;
Status = success.
```

9.12 Example 12

In the structure of the image all the joints are pin joints. Ignoring the weight of the beams, get all the reactions in the joints.

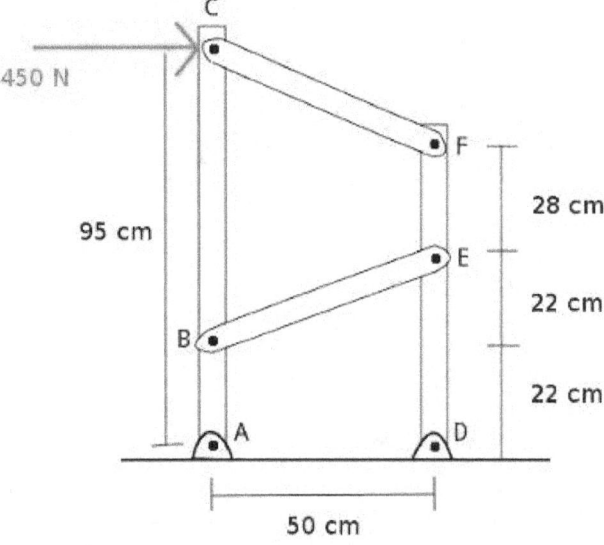

Solution with FisicaLab

Notice that the beams BE and CF are beams of two forces. So, we build the problem with two elements Beam, and the necessary elements Force and Beams of 2 forces, as show below:

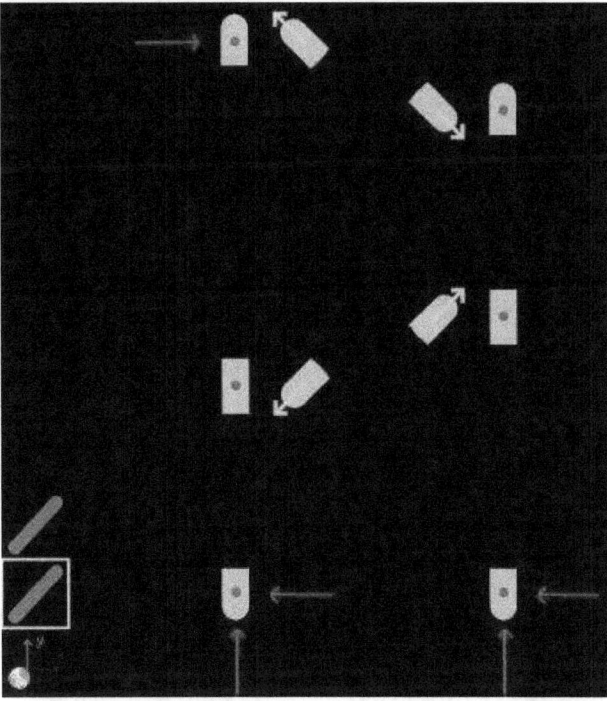

The Stationary reference system set the gravity value to default. Now to the element Beam that represent the ABC beam, we have:

Name	ABC
m	0
lc	0
ang	90

This beam have three elements: A, B and C. To the element that correspond with A:

Name	ABC
l	0

And to the applied Forces, the reactions:

f	RAx

f	RAy

Now to the element that correspond with B:

Name	ABC
l	22 @ cm

And to the element Beam of 2 forces, that correspond with the beam from B to E, the angle is entered as the slope 22/50:

f	BE
ang	23.749

To the element that correspond with C:

Name	ABC
l	95 @ cm

To the element Force, that correspond with the force of 450 N:

f	450

And to the element Beam of 2 forces, that correspond with the beam from C to F, the angle is entered as the slope 23/50:

f	CF
ang	24.702

Now to the element Beam that represent the DEF beam, we have:

Name	DEF
m	0
lc	0
ang	90

This beam have three elements: D, E and F. To the element that correspond with D:

Name	DEF
l	0

And to the applied Forces, the reactions:

f	RDy

f	RDx

To the element that correspond with E:

Name	DEF
l	44 @ cm

And to the element Beam of 2 forces, that correspond with the beam from B to E, the angle is also the slope 22/50:

f	BE
ang	23.749

To the element that correspond with F:

Name   DEF

l   72 @ cm

And to the element Beam of 2 forces, that correspond with the beam from C to F, the angle is also the slope 23/50:

f   CF

ang   24.702

Now click in the icon Solve to get the answer:

```
RDy = 854.981 N ;   RDx = -461.094 N ;
BE = -1295.364 N [tension] ;
CF = 797.558 N [compression] ;   RAy = -854.981 N ;
RAx = 911.094 N ;
Status = success.
```

9.13 Example 13

In the truss of the image, find the support reactions and internal forces in all beams under the load conditions shown.

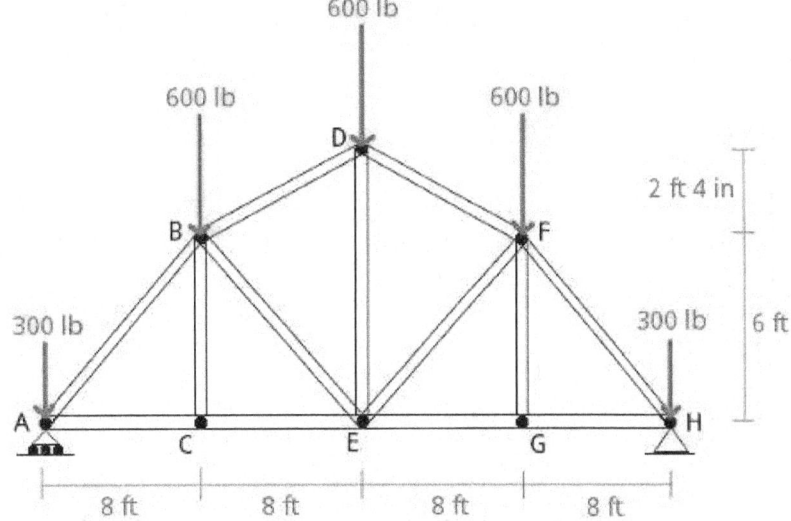

Solution with FisicaLab

Select the group Static and, inside this, the module Static of solids. Erase the content of the chalkboard and select the English system. Then add one element Truss, eight elements Joint of truss, eight elements Force, five forces for the load forces and three for the support reactions (notice that in support A there is only one vertical force), and 26 elements Beam of truss to build the problem as show below:

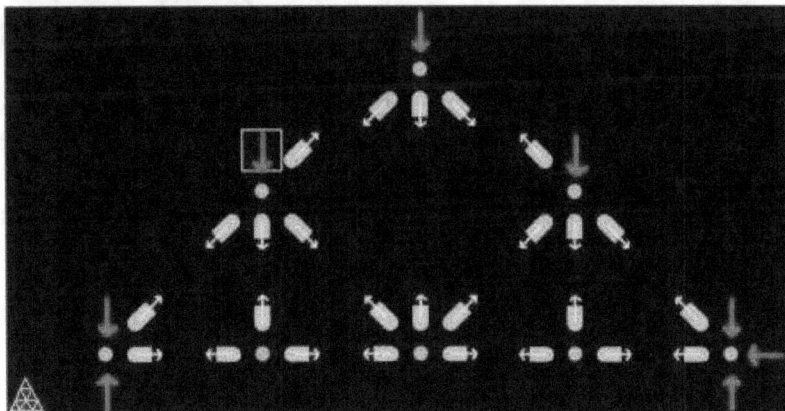

To the element Truss, we use the name AH:

Name   AH

All the elements Joint of truss, will owned by this truss. So, we add each joint to this truss:

Truss   AH

For the two elements Beam of truss that represent the beam from joint A to C, and calling AC the internal force, we have:

t   AC

Now for the two elements Beam of truss that represent the beam from joint A to B, and calling AB the internal force and entering the angle as the slope 6/8, we have:

t   AB

ang   36.870

For the two elements Beam of truss that represent the beam from joint B to C, and calling BC the internal force, we have:

t   BC

For the two elements Beam of truss that represent the beam from joint B to E, and calling BE the internal force, we have:

t   BD

ang   16.260

Here the angle is entered as slope 28/96. The numerator and denominator should be at the same units: 2 ft 4 in = 28 in and 6 ft = 96 in. Also can be used 2.333/6 using the feet unit. Now for the two elements Beam of truss that represent the beam from joint B to E, and calling BE the internal force, we have:

t   BE

ang   36.870

Here the angle is also the slope 6/8. For the two elements Beam of truss that represent the beam from joint C to E, and calling CE the internal force, we have:

t   CE

For the two elements Beam of truss that represent the beam from joint
D to E, and calling DE the internal force, we have:

t   DE

Now for the two elements Beam of truss that represent the beam from
joint D to F, and calling DF the internal force, we have:

t   DF

ang   16.260

Here the angle is entered as slope 28/96. For the two elements Beam of
truss that represent the beam from joint E to F, and calling EF the internal
force, we have:

t   EF

ang   36.870

Here the angle is entered as slope 6/8. Now for the two elements Beam of
truss that represent the beam from joint E to G, and calling EG the internal
force, we have:

t   EG

For the two elements Beam of truss that represent the beam from joint
F to G, and calling FG the internal force, we have:

t   FG

For the two elements Beam of truss that represent the beam from joint
F to H, and calling FH the internal force, we have:

t   FH

ang   36.870

Here the angle is entered as slope 6/8. Now for the two elements Beam of
truss that represent the beam from joint G to H, and calling GH the internal
force, we have:

t   GH

Now to the element Force that represent the reaction in A:

f   RAy

And to the two elements Force that represent the reactions in H:

f   RHy

f   RHx

Now to the elements Force that represent the load forces, three of 600 lb and two of 300 lb:

f   300

f   600

Now click in the icon Solve to get the answer:

```
FG = 0.000 lb [tension] ;   EF = -59.993 lb [compression] ;
DF = -1200.000 lb [compression] ;   RAy = 1200.000 lb ;
FH = -1499.996 lb [compression] ;
AC = 1199.996 lb [tension] ;
AB = -1499.996 lb [compression] ;
GH = 1199.996 lb [tension] ;
RHy = 1200.000 lb ;   RHx = 0.000 lb ;
BC = -0.000 lb [compression] ;
CE = 1199.996 lb [tension] ;
BD = -1200.000 lb [compression] ;
DE = 71.992 lb [tension] ;   BE = -59.993 lb [compression] ;
EG = 1199.996 lb [tension] ;
Status = success.
```

10 Module dynamics of particles

The conversion factors for SI system are:

h	hour
min	minute
hp	horse power
cv	steam horses
T	metric ton
g	gram
slug	slug
km	kilometer
cm	centimeter
mm	millimeter
mi	mile
ft	feet
in	inch
km/h	kilometers per hour
cm/s	centimeters per second
mm/s	millimeters per second
mph	miles per hour
ft/s	feet per second
in/s	inch per second
kt	knot
cm/s2	centimeters per squared second
mm/s2	millimeters per squared second
ft/s2	feet per squared second
in/s2	inch per squared second

And for the English system are:

h	hour
min	minute
hp	horsepower
cv	steam horses
kg	kilogram
g	gram
km	kilometer
m	meter
cm	centimeter
mm	millimeter
mi	mile
in	feet
km/h	kilometers per hour
m/s	meters per second
cm/s	centimeters per second
mm/s	millimeters per second
mph	miles per hour
in/s	inch per second
kt	knot
m/s2	meters per squared second
cm/s2	centimeters per squared second
mm/s2	millimeters per squared second
in/s2	inch per squared second
lb/in	pound per inch

In this module, the forces (elements Force, Friction or Contact) are added to the objects (elements Block, Pulley or Spring) placing the force in one of the cells around the object. Each object (if there isn't in the chalkboard's border) have 8 cells around it. This module have 44 elements, presented below. With a description of each one, its data and its number of equations. Note that some elements have one or two equations, depending of the applied forces, or of the objects that interact with it.

10.1 Stationary reference system

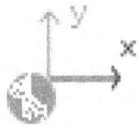

Stationary reference system, with X axis horizontal and positive to the right, and Y axis vertical an positive to upwards.

Equations: None.

Data:

g: Value of the gravity (absolute value). To default FisicaLab write the gravity value to the selected system (9.81 m/s2 to SI, and 32.2 ft/s2 to English system).

t: End time to the problem.

10.2 Mobile

Mobile to collision problems. The angles are measured from the positive X axis, and the positive sense is the opposite of clockwise.

Equations: None.

Data:

Name: Name of the mobile.

m: Mass of the mobile.

vi: Initial velocity, in the direction specified by the initial angle.

angi: Initial angle.

vf: Final velocity, in the direction specified by the final angle.

angf: Final angle.

10.3 Mobile in X/Y

Mobiles with movement along the X and Y axis, respectively.

Equations: None.

Data:
Name: Name of the mobile.

m: Mass of the mobile.

vi: Initial velocity.

vf: Final velocity.

10.4 Block with vertical movement

    Block with vertical movement. Allows only one friction force, vertical or horizontal.

Equations:	3 or 4 (if all the applied forces are vertical the block have 3 equation, otherwise have 4).

Data:

Name:	Name of the block.
m:	Mass block.
a:	Acceleration of the block.
vi:	Initial velocity.
vf:	Final velocity.
d:	Travelled distance.
Relative to:	To default is *sf* meaning that the data is measured from a stationary reference system. But you can set a block whit horizontal movement, if the data is relative to that block.

10.5 Block with horizontal movement

Block with horizontal movement. Allows only one horizontal friction force.

Equations: 4

Data:

Name: Name of the block.

m: Mass of the block.

a: Acceleration of the block.

vi: Initial velocity.

vf: Final velocity.

d: Travelled distance.

Relative   To default is *sf* meaning that the data is
to: measured from a stationary reference sys-
              tem. But you can set a block, if the data is
              relative to that block.

10.6 Block with movement along an inclined plane to the left

The data of this object are referring to a reference system, with the X axis along the plane and positive to upwards. And Y axis perpendicular to the plane and positive to upwards. Allows only one friction force parallel to the plane.

Equations: 4

Data:

Name:	Name of the block.
m:	Mass of the block.
ang:	Angle of the plane, measured from the horizontal.
a:	Acceleration of the block.
vi:	Initial velocity.
vf:	Final velocity.
d:	Travelled distance.
Relative to:	To default is *sf* meaning that the data is measured from a stationary reference system. But you can set a block, if the data is relative to that block.

10.7 Block with movement along an inclined plane to the right

The data of this object are referring to a reference system, with the X axis along the plane and positive to upwards. And Y axis perpendicular to the plane and positive to upwards. Allows only one friction force parallel to the plane.

Equations: 4

Data:

Name: Name of the block.

m: Mass of the block.

ang: Angle of the plane, measured from the horizontal.

a: Acceleration of the block.

vi: Initial velocity.

vf: Final velocity.

d: Travelled distance.

Relative to: To default is *sf* meaning that the data is measured from a stationary reference system. But you can set a block, if the data is relative to that block.

10.8 Pulley

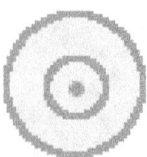

Pulley without mass. Don't allows frictions or contacts.

Equations: 1 or 2 (if all applied forces are vertical, or horizontal, the block have 1 equation, otherwise have 2).

Data:

Name: Name of the pulley (this data is irrelevant).

10.9 Springs

Springs without mass, that satisfies the Hooke's law. *xi* and *xf* are, respectively, the initial and end length that the spring is stretched or compressed. If the spring is stretched the length is positive, if is compressed is negative. Allow one or two forces. If have two forces these must have different sense (stretching or compressing the spring), and the same values or unknown data. Don't allows frictions or contacts. The forces must be in the direction which will move the spring (see examples).

Equations: 1

Data:

k: Spring constant.

xi: Initial length.

xf: Final length.

10.10 Forces

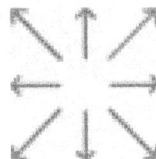

Forces with the direction and sense of the arrow.

Equations: None.

Data:

f   Magnitude of the force.

ang   Positive angle of the force, measured from
                 the horizontal. This entry is displayed only
                 for oblique forces.

10.11 Frictions

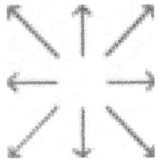

Frictions with the direction and sense of the arrow. The oblique frictions, to be applied to a block on an inclined plane, are assumed parallels to the plane.

Equations: None.

Data:

Normal: Normal to calculate the friction force.

u: Friction coefficient, static or dynamic.

10.12 Frictions between blocks (contacts)

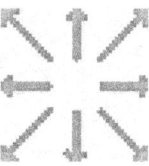

Friction forces between blocks, with the direction and sense of the arrows.

Equations: None.

Data:

Normal: Normal to calculate the friction force.

u: Friction coefficient, static or dynamic.

ang: Positive angle of the friction, measured from the horizontal. This entry is displayed only for oblique frictions.

10.13 Relation between accelerations

a1 =
z*a2

Allows to make a relation between the magnitudes of accelerations measured in different reference systems.

Equations: 1

Data:

a1:   Acceleration of the first block.

a2:   Acceleration of the second block.

z: The relation factor, to default is -1.

10.14 Relative motion

Allows get the acceleration, final velocity and travelled distance of a block in reference to the stationary system. When its data is relative to another block, that making a mobile reference system. The angles are measured from the positive X axis, and the positive sense is the opposite of clockwise.

Equations: 6

Data:

Object: Name of the object that its data is relative.

asf: Magnitude of the acceleration.

ang_asf: Angle of the acceleration vector.

vf_sf: Magnitude of the final velocity.

ang_vfsf: Angle of the final velocity vector.

dsf: Travelled distance of the block.

ang_dsf: Angle of the travelled distance.

10.15 Collision

Collision between mobiles. Allows collision between two Mobile in X, or two Mobile in Y. Or between two Mobile. Any other collision isn't allowed.

Equations: 2 or 4 (if the collision is between two mobiles in x, or two mobiles in Y, have 2 equations. Otherwise 4).

Data:

Object 1:   Name of first mobile.

Object 2:   Name of second mobile.

e: Coefficient of restitution.

angn: Angle of the normal between the objects (when collide). Measured from the positive X axis. The positive sense is the opposite of clockwise.

10.16 Energy

Set the principle of energy and work to 1, 2, 3, or 4 blocks or mobiles.

Equations: 1

Data:

Object 1:   Name of the first block or mobile.

Object 2:   Name of the second block or mobile.

Object 3:   Name of the third block or mobile.

Object 4:   Name of four block or mobile.

W: External work to the system.

10.17 Momentum

Set the principle of momentum and impulse, to one block or mobile.

Equations: 2

Data:

Object: Name of the block or mobile.

Imp: Magnitude of the impulse's vector.

ang: Angle of the impulse's vector. Measured
                from the positive X axis. The positive sense
                is the opposite of clockwise.

fImp: Impulsive force.

10.18 Power

P

It measures the developed power.

Equations: 1

Data:

Object: Name of the block.

P: Power.

11 Examples dynamics of particles

11.1 Example 1

A block of 3 kilograms of mass is above of 30 degrees inclined plane. If it leave the rest and there isn't friction, Which is the velocity of the block after travelled 2 meters? Which is its acceleration?

Solution with FisicaLab

Select the Dynamic group and, inside this, the Particles module. Erase the content of the chalkboard and select the SI system. And add one element Oblique force, one element Block above an inclined plane to the right and one element Stationary reference system, to make the free body diagram of the block. As show the image below:

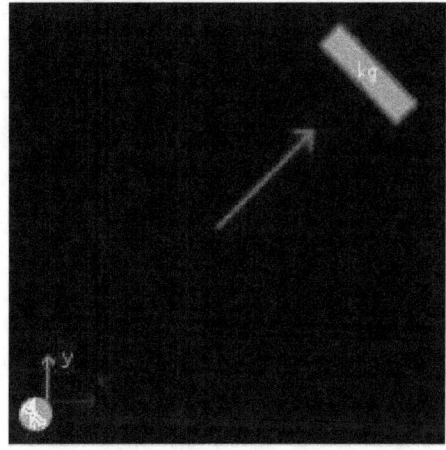

The element Stationary reference system write to default the gravity value, but the time data is unknown:

g	9.81
t	t

Now, to the element Block above an inclined plane to the right, write:

Name	Block
m	3
ang	30
a	a
vi	0
vf	vf

d   -2

Relative to
            sf

The sign of the travelled distance is negative (remember the reference system to these elements). To the element Oblique force write:

f   n

ang   60

If the plane is inclined 30 degrees, the normal force is inclined 60 degrees. Now click in the icon Solve to get the answer:

```
a = -4.905 m/s2 ;   vf = -4.429 m/s ;   t = 0.903 s ;
n = 25.487 N ;
Status = success.
```

The relevant data are vf = -4.426 m/s and a = -4.905 m/s2, both with negative sign (remember the reference system to these elements).

11.2 Example 2

A block of 4 kilograms of mass is above a scales inside an elevator with an acceleration of 1.6 m/s2 to the upwards. Which is the weight that show the scales?

Solution with FisicaLab

Erase the content of the chalkboard and select the SI system. And add one element Force, one element Block with vertical movement and one element Stationary reference system, to make the free body diagram of the block. As show the image below:

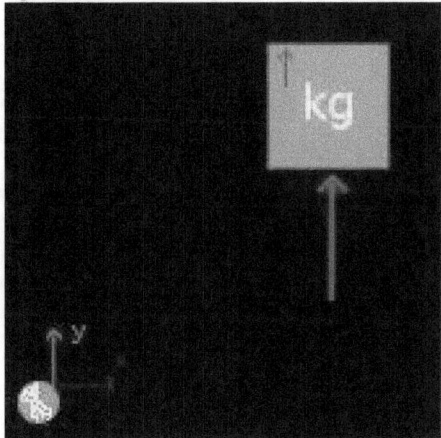

We don't have a time data, but we can assume anytime. Here we assume 1 second. Also, we don't have an initial velocity data, we will assume 0. Then to the element Stationary reference system write:

g	9.81
t	1

The searched wight is the same if we assume other time or initial velocity. To the element Block with vertical movement write:

Name	Block
m	4
a	1.6
vi	0
vf	vf
d	d
Relative to	
	sf

To the element Force, that is the normal force and the searched weight, write:

f   n

Now click in the icon Solve to get the answer:

```
n = 45.640 N ;   vf = 1.600 m/s ;   d = 0.800 m ;
Status = success.
```

The relevant data is n = 45.640 N.

11.3 Example 3

Two blocks are joined with a string as show the image below. If the system leaves the rest, which is the acceleration of the system? Which is the velocity of each block after 0.5 seconds? The travelled distance?

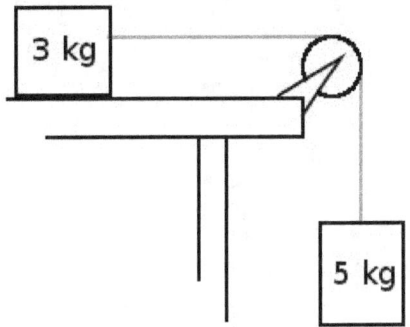

Solution with FisicaLab

Erase the content of the chalkboard and select the SI system. And add one element Stationary reference system, one element Block with vertical movement, one element Block with horizontal movement, and one element Relation between accelerations. And add the necessary elements Force to make the free body diagram of each block. As show the image below:

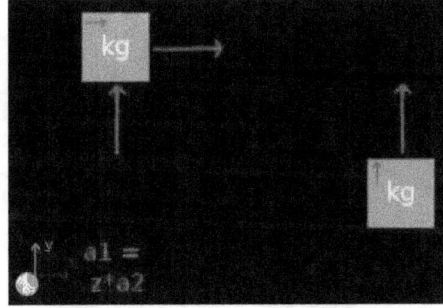

To the element Stationary reference system, write:

g	9.81
t	0.5

To the block that correspond with the block above the table, we call this *Block A*, write:

Name	Block A
m	3
a	a_A
vi	0

vf	vf_A
d	d_A

Relative to
> sf

The name is irrelevant. To its normal force write:

f	n_A

And to the force of the string:

f	t

To the other block, we call this *Block B*, write:

Name	Block B
m	5
a	a_B
vi	0
vf	vf_B
d	d_B

Relative to
> sf

And to the force of the string:

f	t

Now, we need make a relation between the accelerations, that have the same magnitude but different sign (In the block above the table the acceleration is positive because is to the right. But in the other block the acceleration is negative because is to the downwards). We do this with the element Relation between accelerations. In this write:

a1	a_A
a2	a_B
z	-1

Now click in the icon Solve to get the answer:

```
a_B = -6.131 m/s2 ;   vf_B = -3.066 m/s ;   d_B = -0.766 m ;
a_A = 6.131 m/s2 ;   t = 18.394 N ;   vf_A = 3.066 m/s ;
d_A = 0.766 m ;   n_A = 29.430 N ;
Status = success.
```

11.4 Example 4

A block of 2 kilograms of mass is above a spring with a constant of 1100 N/m that is stretched 5 cm, as show the image. The dynamic friction coefficient between the block and the plane is 0.2. If the spring is released suddenly, and the friction force acts from which the block leaves the spring, Which distance travelled the block above the plane? Without the 5 cm that the spring is stretched. How much energy is lost through friction?

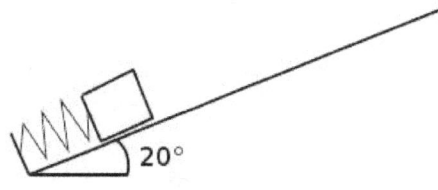

Solution with FisicaLab

Erase the content of the chalkboard and select the SI system. And add one element Stationary reference system, one element Block above an inclined plane to the left and one element Spring. And add the necessary elements Force to make the free body diagram of block and the spring. As show the image below:

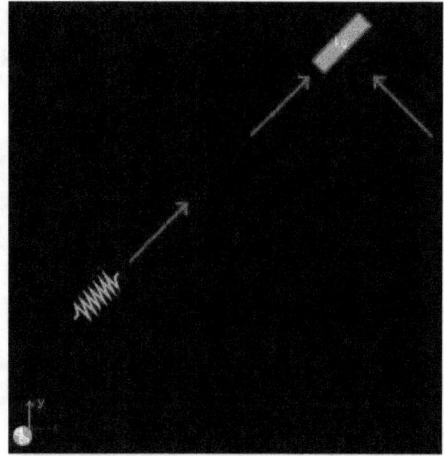

Remember that the forces applied to a spring must be in the direction which will move the spring.

We need solve the problem in two steps. First we need know the velocity of the block when leaves the spring. And after applied a friction force to know the travelled distance. To first step, in the element Stationary reference system write:

g	9.81
t	t

Because time data is unknown. To the spring:

Name	Spring
k	1100
xi	-5 @ cm
xf	0

Here the name is irrelevant. The initial position is negative because the spring is stretched. And final position is 0. To the force applied to the spring, that is an unknown data, write:

f	f
ang	20

Here the angle is irrelevant. To the block:

Name	Block
m	2
ang	20
a	a
vi	0
vf	vf
d	5 @ cm
Relative to	sf

The travelled distance is the longitude that the spring is stretched. To the force that correspond with the normal, that is a 70 degrees, write:

f	n
ang	70

And to the force applied to the spring, write:

f	n
ang	20

Now click in the icon Solve to get the answer:

```
n = 18.437 N ;   f = 27.500 N ;   t = 0.098 s ;
a = 10.395 m/s2 ;   vf = 1.020 m/s ;
Status = success.
```

The acceleration data is the average acceleration, because the force isn't constant. The force is the average force too.

Now, to the second step, delete the spring and its forces. And add one element Energy and one element Friction, as show the image below:

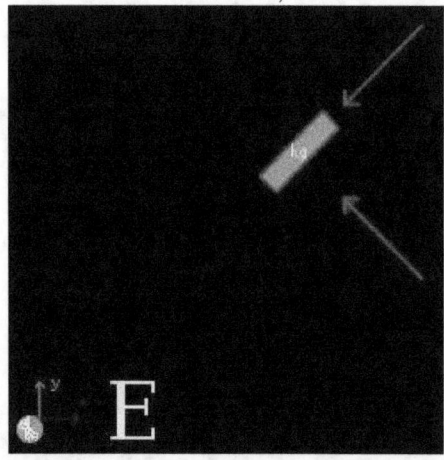

The data in the element Stationary reference system is the same:

g	9.81
t	t

Now, to the block write:

Name	Block
m	2
ang	20
a	a
vi	1.020
vf	0
d	d
Relative to	
	sf

Now the name to this element is important. The initial velocity is the velocity calculated in the first step. And the end velocity is 0. Also, the distance travelled is now an unknown data. To the normal force the data is the same:

f	n
ang	70

To the element Friction write:

Normal	n
u	0.2

Where n is the normal to calculate the friction force. And to the element Energy:

Object 1	Block
Object 2	0
Object 3	0
Object 4	0
W	Work

Now click in the icon Solve to get the answer:

```
n = 18.437 N ;   a = -5.199 m/s2 ;   d = 0.100 m ;
t = 0.196 s ;   Work = -0.369 J ;
Status = success.

WARNING: Check that the sense of friction forces is
opposite of the movement sense (when correspond).
```

The lost energy is 0.369 Joules, and the travelled distance is 0.100 meters or 10 cm.

11.5 Example 5

A spring with a constant of 1300 N/m, is stretched 8 cm, and is between two blocks as show the image. One block have a mass of 0.5 kg and the other have 0.7 kg. If the spring is released suddenly, and there isn't friction, Which is the velocity of each block when these leaves the spring?

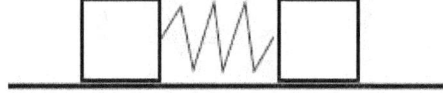

Solution with FisicaLab

Erase the content of the chalkboard and select the SI system. And add one element Stationary reference system, two elements Block with horizontal movement, one element Spring, and one element Energy. And add the necessary elements Force to make the free body diagram of each block and the spring. As show the image below:

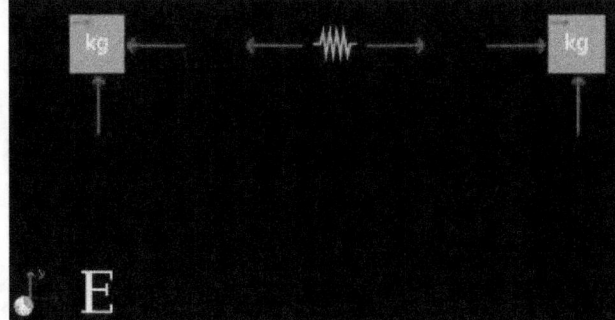

To the element Stationary reference system the time is an unknown data:

g	9.81
t	t

To the block at the left, that we assume is the block of 0.5 kg, write:

Name	A
m	0.5
a	a_A
vi	0
vf	vf_A
d	d_A
Relative to	
	sf

To the normal force of this block, write:

f   n_A

To the block at the right, of 0.7 kg, write:

Name	B
m	0.7
a	a_B
vi	0
vf	vf_B
d	d_B
Relative to	
	sf

To its normal force:

f   n_B

To each force applied on the blocks to the spring, write:

f   f

To each force in the spring, write:

f   f

The same unknown data that the forces applied to the blocks. Now, to the element Spring:

Name	Spring
k	1300
xi	-8 @ cm
xf	0

And in the element Energy:

Object 1   A

Object 2   B

Object 3   Spring

Object 4   0

W   0

The work is 0, because there isn't friction. Now click in the icon Solve to get the answer:

```
t = 0.030 s ;   f = 52.000 N ;   n_B = 6.867 N ;
a_A = -104.000 m/s2 ;   vf_A = -3.116 m/s ;
d_A = -0.047 m ;   a_B = 74.286 m/s2 ; vf_B = 2.225 m/s ;
d_B = 0.033 m ;   n_A = 4.905 N ;
Status = success.
```

The relevant data are vf_A = -3.116 m/s and vf_B = 2.225 m/s.

11.6 Example 6

A mobile A with a mass of 3 kg and velocity of 50 m/s, is moving in the positive X direction. A second mobile B with a mass of 5 kg, and velocity of 75 m/s, is moving in the positive X direction too. If the A mobile impact the B mobile and if the impact is elastic, which is the velocity of each mobile after the impact?

Solution with FisicaLab

Erase the content of the chalkboard and select the SI system. And add one element Collision, and two elements Mobile in X. As show the image below:

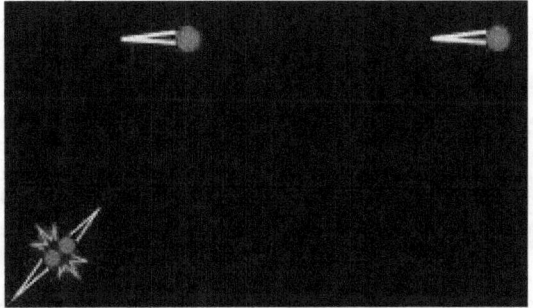

To the element Mobile in X at the left, that we assume is the mobile A, write:

Name	A
m	3
vxi	50
vxf	vf_A

To the other mobile, the B mobile, write:

Name	B
m	5
vxi	75
vxf	vf_B

To the element Collision write:

Object 1	A
Object 2	B
e	1

angn   0

 e is 1 because the collision is elastic, and angn is 0 because the impact is central. Now click in the icon Solve to get the answer:

```
vf_A = 81.250 m/s ;   vf_B = 56.250 m/s ;
Status = success.
```

11.7 Example 7

An A mobile of 4 kg and a B mobile of 3 kg, with velocities of 80 m/s and 110 m/s respectively, have the directions that show the image. These mobiles impact with the normal in the image. If the coefficient of restitution is 0.7, which are the velocity and direction of each mobile after the impact?

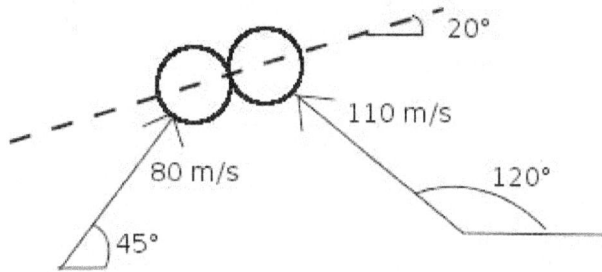

Solution with FisicaLab

Erase the content of the chalkboard and select the SI system. And add one element Collision, and two elements Mobile. As show the image below:

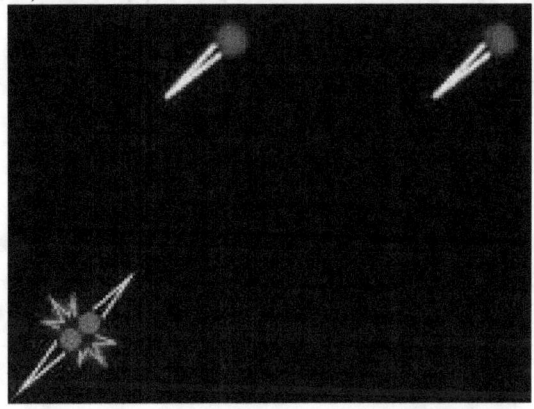

To the element Mobile at the left, that we assume is the A mobile, write:

Name	A
m	4
vi	80
angi	45
vf	vf_A
angf	ang_A

To the other mobile, the B mobile:

Name	B
m	3

vi   110

angi   120

vf   vf_B

angf   ang_B

And to the element Collision:

Object 1   A

Object 2   B

e   0.7

angn   20

Now click in the icon Solve to get the answer:

```
vf_A = 34.297 m/s ;   ang_A = 100.326 degrees ;
vf_B = 128.916 m/s ;   ang_B = 77.172 degrees ;
Status = success.
```

11.8 Example 8

A block A is above a block B, as show the image. Both blocks are in rest. If the block A have a mass of 1.7 kg, block B have 4 kg, and there isn't friction, Which are the accelerations of the blocks when these leaves the rest? Which are the velocities after 0.2 seconds?

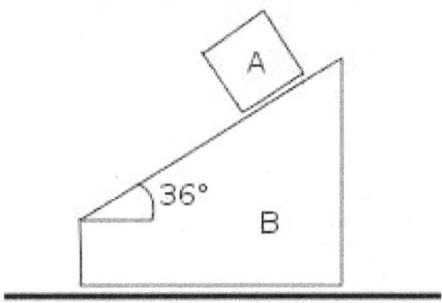

Solution with FisicaLab

Erase the content of the chalkboard and select the SI system. And add one element Stationary reference system, one element Block with horizontal movement, one element Block above an inclined plane to the left, and one element Relative motion. And add the necessary elements Force to make the free body diagram of each block. As show the image below:

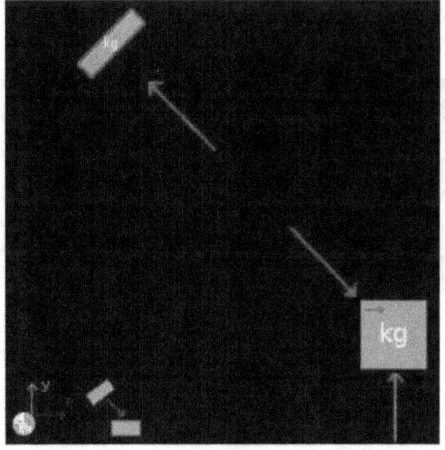

To the element Stationary reference system, write:

g	9.81
t	0.2

To the element Block above an inclined plane to the left, the block A, write:

Name	A
m	1.7

ang	36
a	aA
vi	0
vf	vfA
d	dA
Relative to	
	B

Note that the movement of this block is relative to the block B. To its normal force:

f	nA
ang	54

To the element Block with horizontal movement, the block B, write:

Name	B
m	4
a	aB
vi	0
vf	vfB
d	dB
Relative to	
	sf

To its normal force:

f	nB

To the force applied to the block A:

f	nA
ang	54

The element Relative motion is necessary to get the acceleration, velocity and travelled distance of the block A, in reference to the stationary system. All these are unknown data, then write:

Object	A
asf	asfA
ang_asf	ang_asfA
vfsf	vfsfA
ang_vfsf	ang_vfsfA
dsf	dsfA
ang_dsf	ang_dsfA

Now click in the icon Solve to get the answer:

```
nA = 11.765 N ;   nB = 48.758 N ;   aA = -7.165 m/s2 ;
vfA = -1.433 m/s ;   dA = -0.143 m ;   aB = 1.729 m/s2 ;
vfB = 0.346 m/s ;   dB = 0.035 m ;   asfA = 5.855 m/s2 ;
ang_asfA = 225.994 degrees ;   vfsfA = 1.171 m/s ;
ang_vfsfA = 225.994 degrees ; dsfA = 0.117 m ;
ang_dsfA = 225.994 degrees ;
Status = success.
```

11.9 Example 9

A baseball ball, of 0.22 kg, is launched with a velocity of 32 m/s horizontally. A batter hits the ball, and this take a velocity of 61 m/s at 40 degrees above the horizontal. Which is the momentum applied to the ball?

Solution with FisicaLab

Erase the content of the chalkboard and select the SI system. And add one element Stationary reference system, one element Mobile, and one element Momentum. As show the image below:

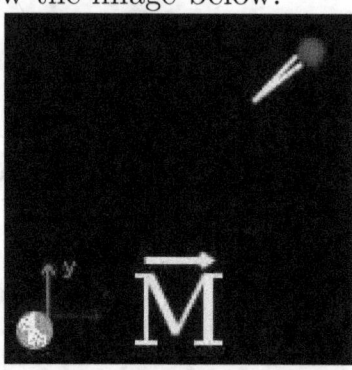

We don't have a time data to the impact (the duration of the impact), but we assume 0.01 seconds, a reasonable time. This time is necessary to calculate the impulsive force in the element Momentum. Then, to the element Stationary reference system, write:

g	9.81
t	0.01

We assume that the ball is, initially, moving in the direction of negative X. Then in the element Mobile, we call this *Ball*, write:

Name	Ball
m	0.22
vi	32
angi	180
vf	61
angf	40

To the element Momentum:

Object	Ball

Imp   imp

ang   ang

fImp   f

Now click in the icon Solve to get the answer:

```
imp = 9.215 N*s ;   ang = 69.412 degrees ;   f = 921.472 N ;
Status = success.
```

11.10 Example 10

A box of 80 kg is rising to a crane, in 3 minutes, at the flat roof of a building of 50 meters. Which is the developed power to the crane?

Solution with FisicaLab

Erase the content of the chalkboard and select the SI system. And add one element Stationary reference system, one element Block with vertical movement, one element Force applied to the block, and one element Power. As show the image below:

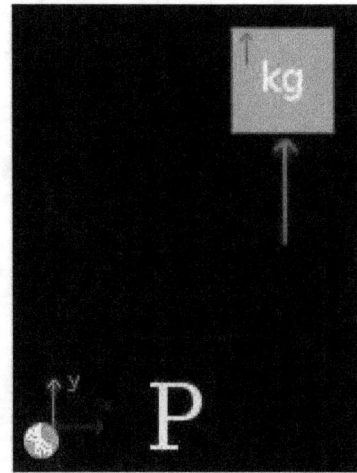

To the element Stationary reference system write:

g	9.81
t	3 @ min

To the block, we assume that this leave the rest and we call this *Box*, write:

Name	Box
m	80
a	a
vi	0
vf	vf
d	50
Relative to	
	sf

To the element Force:

f   f

To the element Power:

Object   Box

P   Power

Now click in the icon Solve to get the answer:

```
f = 785.047 N ;   a = 0.003 m/s2 ;   vf = 0.556 m/s ;
Power = 218.069 W ;
Status = success.
```

NOTE: *You can also calculate the developed power when the box is rising with a constant velocity. To do this, write 0 in the acceleration, and initial velocity and final velocity as unknown data.*

11.11 Example 11

An A block of 23 kg, is above a B block of 17 kg, as show the image. If the friction coefficient in all surfaces is 0.18, and F force is of 235 Newtons, which is the acceleration of the block A? Which is the stress in the string?

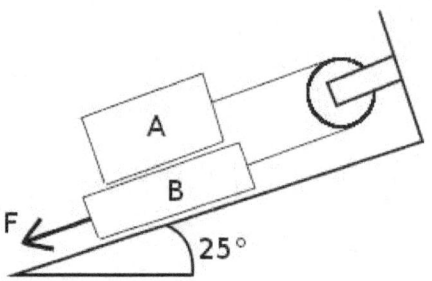

Solution with FisicaLab

Erase the content of the chalkboard and select the SI system. And add the necessary elements to make the free body diagram of each block. As show the image below:

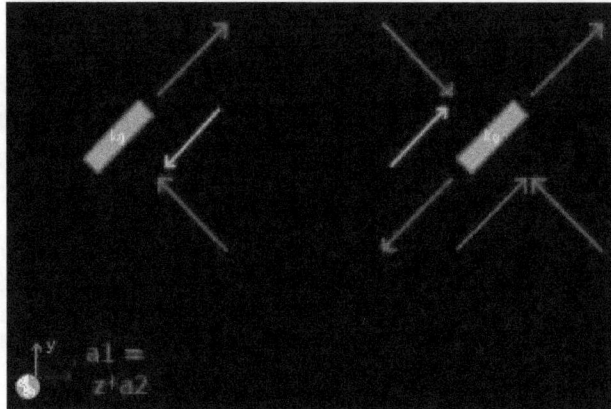

We assume that the system leaves the rest, and assume a time of 0.5 seconds. Then to the element Stationary reference system write:

g	9.81
t	0.5

To the block that correspond with the A block, write:

Name	A
m	23
ang	25
a	aA
vi	0

vf	vfA
d	dA
Relative to	
	sf

To its normal:

f	nA
ang	65

To the element Friction between blocks write:

N	nA
u	0.18
ang	25

To the force that correspond with force applied to the string, write:

f	t
ang	25

Now, to the B block, write:

Name	B
m	17
ang	25
a	aB
vi	0
vf	vfB
d	dB
Relative to	
	sf

To its normal:

f	nB

ang   65

To the friction with the plane:

N   nB
u   0.18

To the friction with A block, the element Friction between blocks, write:

N   nA
u   0.18
ang   25

To the applied force of 235 Newtons, write:

f   235
ang   25

To the force that correspond with force applied to the string, write:

f   t
ang   25

To the force applied to the A block (the normal reaction):

f   nA
ang   65

And to the relation between the accelerations:

a1   aA
a2   aB
z   -1

Now click in the icon Solve to get the answer:

```
nB = 355.635 N ;   t = 173.848 N ;   nA = 204.490 N ;
aA = 1.812 m/s2 ;   aB = -1.812 m/s2 ;   vfB = -0.906 m/s ;
```

```
dB = -0.227 m ;   vfA = 0.906 m/s ;   dA = 0.227 m ;
Status = success.
```

WARNING: Check that the sense of friction forces is opposite of the movement sense (when correspond).

11.12 Example 12

A cannon of 1000 kg, shoot a projectile of 9 kg with a velocity of 500 m/s at 30 degrees above the horizontal. If the cannon is above a horizontal surface and is free to move, and is a rigid body, which is the velocity of the cannon after the shoot? Which is the normal applied to the floor? The shoot last 7 milliseconds.

Solution with FisicaLab

Erase the content of the chalkboard and select the SI system. And add the elements as show the image below:

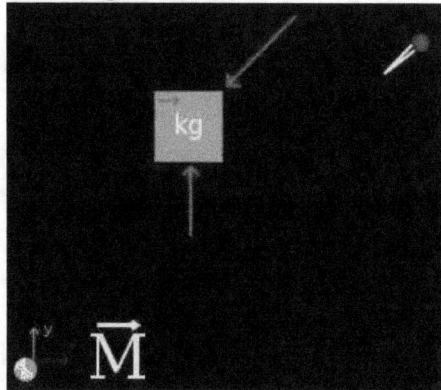

To the element Stationary reference system write:

g	9.81
t	0.007

To the block:

Name	Cannon
m	1000
a	a
vi	0
vf	vf
d	d
Relative to	
	sf

To its normal force:

f	n

To the impulsive force produced to the shoot (the oblique force), write:

f	f
ang	30

To the element Mobile, the projectile that we call Bala (projectile), write:

Name	Projectile
m	9
vi	0
angi	0
vf	500
angf	30

Now through the element Momentum we apply the impulsive force to the object Projectile:

Object	Projectile
Imp	imp
ang	ang
fImp	f

Now click in the icon Solve to get the answer:

```
n = 331238.571 N ;   a = -556.731 m/s2 ;   vf = -3.897 m/s ;
d = -0.014 m ;   f = 642857.143 N ;   imp = 4500.000 N*s ;
ang = 30.000 degrees ;
Status = success.
```

12 Module circular dynamics of particles

The conversion factors for SI system are:

mm	millimeter
cm	centimeter
km	kilometer
in	inch
ft	feet
mi	mile
mm/s	millimeters per second
mm/s2	millimeters per squared second
cm/s	centimeters per second
cm/s2	centimeters per squared second
km/h	kilometers per hour
in/s	inch per second
in/s2	inch per squared second
ft/s	feet per second
ft/s2	feet per squared second
mph	miles per hour
kt	knot
min	minute
h	hour
rad	radian
rpm	revolutions per minute
g	gram
T	metric ton
slug	slug
cv	steam horses
hp	horsepower
N*cm	Newton centimeter

And for the English system are:

in	inch
in/s	inch per second
in/s2	inch per squared second
mm	millimeter
mm/s	millimeters per second
mm/s2	millimeters per squared second
cm	centimeter
cm/s	centimeters per second
cm/s2	centimeters per squared second
m	meter
m/s	meters per second
m/s2	meters per squared second
km	kilometer
km/h	kilometers per second
kt	knot
mi	mile
mph	miles per hour
min	minute
h	hour
rad	radian
rpm	revolutions per minute
g	gram
kg	kilogram
cv	steam horses
hp	horsepower
lb/in	pounds per inch

In this module the forces (elements Force or Friction) are added to the objects (elements Mobile and Spring) placing the force in one of the cell around the object. Each object (if there isn't in the chalkboard's border) have 8 cells around it. This module have 42 elements described below, its data, equations, and what elements and combinations allows.

In this module the elements Object in rest, Mobile with linear movement, Mobile with circular movement, Mobile with polar circular movement, Spring and Center of rotation are used to set the initial and final state, separately, of the problem. Then both states are related through the elements Energy, Angular Momentum, Linear Momentum and/or Power. The element Mobile with perpendicular circular movement can be used to set/know additional conditions (see examples).

Note: *Items Energy, Angular Momentum, Linear Momentum and Power, admit only certain combinations of objects. And although in general the user should not worry about this, at the end of this section we provides as reference a complete list of supported combinations for each of these elements.*

12.1 Stationary reference system

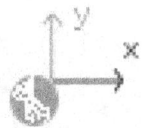

Stationary reference system, with X axis horizontal and positive to the right, and Y axis vertical an positive to upwards.

Equations: None.

Data:

g: Value of the gravity (absolute value). To default write the gravity value to the selected system (9.81 m/s2 to SI, and 32.2 ft/s2 to English system).

t: Time of the problem, to be used in data of power and momentum. To default writes 1, to avoid a power undefined data.

12.2 Object in rest

An object in rest with a Y coordinate above the horizontal reference.

Equations: None.

Data:

Name: Object name.

m: Mass.

y: Vertical coordinate.

12.3 Mobile with linear movement

Mobile with linear movement. Allows apply forces and frictions.

Equations: 2 if there are forces, in other case 0.

Data:

Name: Object name.

m: Mass.

v: Velocity of the mobile.

x: X coordinate of the mobile.

y: Y coordinate of the mobile.

ang: Angle of the velocity vector. Measured from the positive X axis, the positive sense is the opposite of clockwise.

a: Acceleration of the mobile.

12.4 Mobile with circular movement

Mobile with circular movement. Allows only one force (not friction), that represent the centripetal force.

Equations: 1 if there is present a force, in other case 0.

Data:

Name: Object name.

m: Mass.

vt: Tangential velocity of the mobile. A positive value mean that the mobile tour in the opposite sense of clockwise.

r: Turning radius.

y: Y coordinate of the mobile.

12.5 Mobile with polar circular movement

Mobile with circular motion described by polar coordinates. A positive value to tangential velocity, mean that the mobile tour in the opposite sense of clockwise. And a positive value to radial velocity, indicates that the mobile moves away from the center of rotation. Allows only one force (not friction), that represent a radial force.

Equations: 1 if there is present a force, in other case 0.

Data:

Name: Object name.

m: Mass.

vt: Tangential velocity of the mobile.

r: Turning radius.

y: Y coordinate of the mobile.

vr: Radial velocity of the mobile. If this data is unknown, the result will be the positive value.

ar: Radial acceleration of the mobile. If this data is unknown, the result will be the positive value.

12.6 Mobile with perpendicular circular movement

Mobile with circular movement in an horizontal plane. This is, perpendicular to the chalkboard's plane. Allows apply forces and frictions.

Equations: 3

Data:

Name: Object name.

m: Mass.

vt: Tangential velocity of the mobile.

r: Turning radius.

at: Tangential acceleration.

Ft: Applied tangential force.

C: Location of the center. By defaults is ">", meaning that the center is at right. While "<" mean that the center is at left.

12.7 Angular velocity

Element to measure the angular velocity of a mobile with circular motion or polar circular motion.

Equations: 1

Data:

Object: Name of the mobile with circular motion or polar circular motion.

vang: Angular velocity. A positive value mean that the mobile tour in the opposite sense of clockwise.

12.8 Centripetal acceleration

Element to measure the centripetal acceleration of a mobile with circular motion or polar circular motion.

Equations: 1

Data:

Object: Name of the mobile with circular motion or polar circular motion.

ac: Centripetal acceleration.

12.9 Angular acceleration

Element to measure the angular acceleration of a mobile with perpendicular circular motion.

Equations: 1

Data:

Object: Name of the mobile with perpendicular circular motion.

aang: Angular acceleration.

12.10 Energy

E

Element to set the principle of the energy and work for two objects or systems. In case of problems involving an element Center, it is assumed that events occur in a horizontal plane, so the gravitational potential energy is not taken into account. Nor does it take into account for the case of the explosion of a projectile. Since it is considered the state an instant before the explosion and the state an instant after it.

Equations: 1

Data:

System i: A System or Object that represent the initial state.

System f: A System or Object that represent the final state.

W: The work involved in changing state.

12.11 Angular momentum

Element to set the principle of angular momentum and angular impulse for two objects or systems.

Equations: 1

Data:

System i: A System or Object that represent the initial state.

System f: A System or Object that represent the final state.

M: The moment involved in changing state. Will be used the time specified in Stationary reference system.

12.12 Linear momentum

Element to set the principle of linear momentum and impulse for two objects or systems.

Equations: 2

Data:

System i: A System or Object that represent the initial state.

System f: A System or Object that represent the final state.

Fx: Horizontal force involved in changing state. Will be used the time specified in Stationary reference system.

Fy: Vertical force involved in changing state. Will be used the time specified in Stationary reference system.

12.13 Power

Element to measure the developed/applied power in changing state. In case of problems involving an element Center, it is assumed that events occur in a horizontal plane, so the gravitational potential energy is not taken into account. Nor does it take into account for the case of the explosion of a projectile. Since it is considered the state an instant before the explosion and the state an instant after it.

Equations: 1

Data:

System i: A System or Object that represent the initial state.

System f: A System or Object that represent the final state.

P: Power.

12.14 Initial System

An initial system composed of several objects. Allows elements Object in rest, Mobile with linear movement, Mobile with circular movement, Mobile with polar circular movement, Center and Springs.

Equations: None.

Data:

Name: Name of the system.

Object 1:   Object part of the system.

Object 2:   Object part of the system.

Object 3:   Object part of the system.

Object 4:   Object part of the system.

12.15 Final System

A final system composed of several objects. Allows elements Object in rest, Mobile with linear movement, Mobile with circular movement, Mobile with polar circular movement, Center and Springs.

Equations: None.

Data:

Name: Name of the system.

Object 1:   Object part of the system.

Object 2:   Object part of the system.

Object 3:   Object part of the system.

Object 4:   Object part of the system.

12.16 Center of rotation

Center of rotation, to be used in a system of mobiles with circular movement that tour around this.

Equations: None.

Data:

Name: Name of the center of rotation.

v: The velocity of the center.

ang: Angle of the velocity vector. Measured from the positive X axis, the positive sense is the opposite of clockwise.

x: X coordinate of the center.

y: Y coordinate of the center.

12.17 Springs

Springs without mass, that satisfies the Hooke's law. d is the length that the spring is stretched or compressed. If the spring is stretched d is positive, if is compressed is negative. Allow one or two forces (not frictions). If have two forces these must have different sense (stretching or compressing the spring), and the same values or unknown data.

Equations: 1.

Data:

k: Spring constant.

d: Length that the spring is stretched or compressed.

12.18 Forces

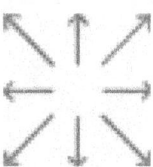

Forces with the direction and sense of the arrow.

Equations: None.

Data:

f: Magnitude of the force.

ang: Positive angle of the force, measured from the horizontal. This entry is displayed only for oblique forces.

12.19 Frictions

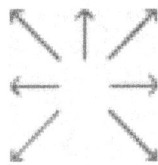

Frictions with the direction and sense of the arrow.

Equations: None.

Data:

f: Magnitude of the force.

u: Static coefficient of friction.

ang: Positive angle of the force, measured from
                the horizontal. This entry is displayed only
                for oblique forces.

12.20 Angles

$$\theta 1 + \theta 2 = 90°$$

To relate two angles that must be complementary. Both angles must be
unknowns.

Equations: 1

Data:

ang1:   One angle.

ang2:   Other angle.

12.21 Moment of a force or couple of forces

Element to relate the moment of a force with the magnitude of the force and the moment arm. Or to relate the moment of a couple of forces with the magnitude of one of the forces and the arm of the couple.

Equations: 1

Data:

M: Moment, a positive value correspond with the opposite sense of clockwise.

f: Magnitude of the force, or the magnitude of one of the forces in the couple.

d: The moment arm, or the arm of the couple.

12.22 Total acceleration (Triangle of accelerations)

Element to measure the vector of the total acceleration of a mobile with perpendicular circular motion.

Equations: 2

Data:

Object: Name of the mobile.

atot: Total acceleration.

ang: Angle of the vector total acceleration, measured from the radius vector. The positive sense is the opposite of clockwise.

12.23 Maximum acceleration

Set the total acceleration, for a mobile with perpendicular circular motion, to the caused by the maximum friction. This to the case when the friction force produces an tangential acceleration and the centripetal acceleration (to the case when the friction force produces only the centripetal acceleration, just apply a friction force to the mobile).

Equations: 1

Data:

Object: Name of the mobile.

u: Static coefficient of friction.

12.24 Inertia

Element to measure the effective radius of rotation and total mass of a system with two or three elements mobile with circular movement.

Equations: 2

Data:

System: Name of the system.

m: Total mass of the system.

r: Effective radius of rotation.

12.25 Absolute velocity

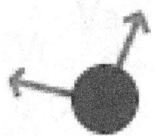

Element to measure the absolute velocity of a mobile with circular movement that turns around an element center.

Equations: 2

Data:

Object: Name of the mobile with circular movement.

angR: Angle of rotation of the mobile, measured from the positive X axis. The positive sense is the opposite of clockwise.

Center   Name of the center.

v: Absolute velocity of the mobile.

ang: Angle of the vector absolute velocity, measured from the positive X axis. The positive sense is the opposite of clockwise.

12.26 Sine of angle

$$\sin \theta = y/r$$

Element to define the sine of an angle as from the radius of a circular path and the height of it above/below the horizontal. The sign of the height is not taken into account, only its absolute value is used.

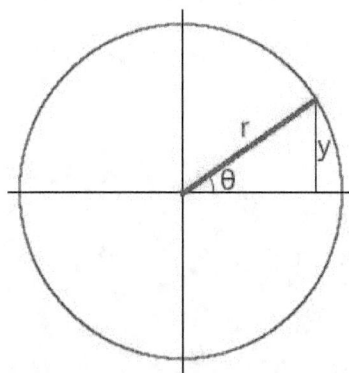

Equations: 1

Data:

y: Height.

r: Turning radius.

ang: Angle.

12.27 Supported combinations of elements Energy, Angular Momentum, Linear Momentum and Power

The following abbreviations are used:

OR   Object in rest.

ML   Mobile with linear movement.

MC   Mobile with circular movement.

MP   Mobile with polar circular movement.

C   Center of rotation.

S   Spring.

Energy and Power elements support the following combinations:

System i	System f
OR	OR
OR	ML
OR	MC
ML	OR
ML	ML
ML	MC
MC	OR
MC	ML
MC	MC
MC	MP
MP	MC
MP	MP
ML	Final system with 2 ML

ML	Final system with 3 ML
ML	Final system with 4 ML
ML	Final system with 2 MC and 1 C
ML	Final system with 2 MC, 1 ML and 1 C
Initial system with 2 MC	Final system with 2 MC
Initial system with 3 MC	Final system with 3 MC
Initial system with 1 MC and 1 S	Final system with 1 MC and 1 S
Initial system with 1 ML and 1 S	Final system with 1 ML and 1 S
Initial system with 1 ML and 2 S	Final system with 1 ML and 2 S
Initial system with 1 ML and 3 S	Final system with 1 ML and 3 S
Initial system with 2 MC and 1 C	Final system with 2 ML
Initial system with 3 MC and 1 C	Final system with 3 ML
Initial system with 3 MC and 1 C	Final system with 2 MC, 1 ML and 1 C
Initial system with 1 MC and 1 S	Final system with 1 OR and 1 S
Initial system with 1 OR and 1 S	Final system with 1 MC and 1 S
Initial system with 1 ML and 1 S	Final system with 1 OR and 1 S
Initial system with 1 OR and 1 S	Final system with 1 ML and 1 S
Initial system with 1 ML and 2 S	Final system with 1 OR and 2 S
Initial system with 1 OR and 2 S	Final system with 1 ML and 2 S
Initial system with 1 ML and 3 S	Final system with 1 OR and 3 S
Initial system with 1 OR and 3 S	Final system with 1 ML and 3 S
Initial system with 1 MC and 1 S	Final system with 1 MP and 1 S
Initial system with 1 MP and 1 S	Final system with 1 MC and 1 S

Initial system with 1 MP and 1 S   Final system with 1 MP and 1 S

The element Angular Momentum supports the following combinations:

System i	System f
MC	MC
MC	MP
MP	MC
MP	MP
ML	Final system with 2 ML
ML	Final system with 3 ML
ML	Final system with 4 ML
ML	Final system with 2 MC and 1 C
ML	Final system with 2 MC, 1 ML and 1 C
Initial system with 2 MC	Final system with 2 MC
Initial system with 3 MC	Final system with 3 MC
Initial system with 2 MC and 1 C	Final system with 2 ML
Initial system with 3 MC and 1 C	Final system with 3 ML
Initial system with 3 MC and 1 C	Final system with 2 MC, 1 ML and 1 C

The element Linear Momentum supports the following combinations:

System i	System f
ML	ML
ML	Final system with 2 ML
ML	Final system with 3 ML
ML	Final system with 4 ML
ML	Final system with 2 MC and 1 C
ML	Final system with 2 MC, 1 ML and 1 C
Initial system with 2 MC and 1 C	Final system with 2 ML
Initial system with 3 MC and 1 C	Final system with 3 ML
Initial system with 3 MC and 1 C	Final system with 2 MC, 1 ML and 1 C

13 Examples circular dynamics of particles

13.1 Example 1

The curve of a road section has a steepness of 18 degrees and a curvature radius of 230 meters. If the coefficient of static friction between the tires of a car and the road is 0.33, find: a) The maximum speed at which the car can make the turn, b) The speed for which there is no friction force between the tires and the road and, c) The minimum speed at which the car can make the turn.

Solution with FisicaLab

Select the Dynamics group and, inside this, the Circular module. Erase the content of the chalkboard and select the SI system. And add one element Stationary reference system, one element Mobile with perpendicular circular movement, one element Force and one element Friction. As show the image below:

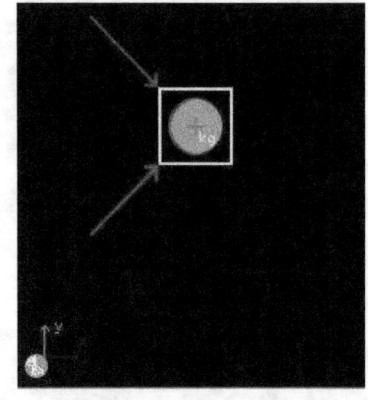

The element Stationary reference system write to default the gravity value. And the time is irrelevant for this problem. Now to the element Mobile with perpendicular circular movement, that represent the car, and since the car's mass is not provided (it is irrelevant for the requested data), we write 1 kg, and we add the conversion factor km/h to the velocity. Also we write as unknown the tangential force, even when tangential acceleration is 0, to satisfy the requirement of number of unknowns. And we let the center of rotation placed to the right, that corresponds with the applied forces:

Name	0
m	1
vt	vt @ km/h
r	230

at	0
Ft	Ft
C	>

For the element Force, which represents the normal, and taking into account that the steepness of the curve is 18 degrees, we have:

f	normal
ang	72

And to the friction force:

N	normal
u	0.33
ang	18

Now click in the icon Solve to get the answer:

```
normal = 11.554 N ;   vt = 146.462 km/h ;   Ft = 0.000 N ;
Status = success.
```

For part b, first remove the friction element, since there should be no friction force:

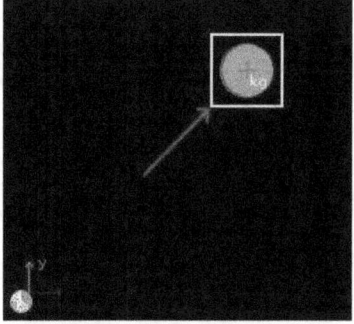

And click in the icon Solve to get the answer:

```
normal = 10.315 N ;   vt = 97.474 km/h ;   Ft = 0.000 N ;
Status = success.
```

Now to part c, add again a friction force:

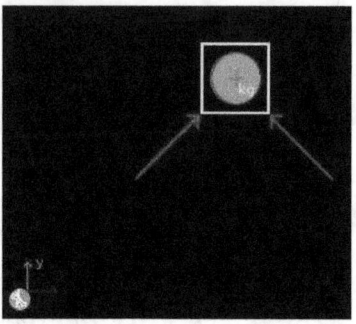

With data:

N	normal
u	0.33
ang	18

And click again in the icon Solve to get the answer:

```
normal = 9.316 N ;  vt = 0.503 km/h ;  Ft = 0.000 N ;
Status = the iteration has not converged yet.
```

What does this mean? That there is no minimum speed. The given coefficient of friction is enough for park the car on the curve without slipping.

13.2 Example 2

A small block A is on a table that rotates from rest with a constant tangential acceleration. If the coefficient of static friction between the block and the turntable is 0.57, find the minimum interval of time for the block to reach a speed of 1.2 m/s without slip.

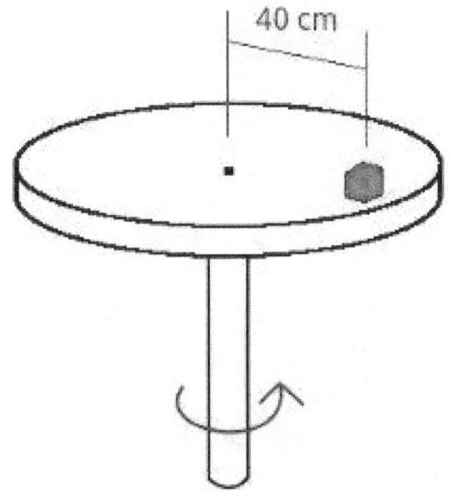

Solution with FisicaLab

Select the Dynamics group and, inside this, the Circular module. Erase the content of the chalkboard and select the SI system. And add one element Stationary reference system, two elements Mobile with circular movement, one element Mobile with perpendicular circular movement, two element Force, one element Angular momentum, one element Moment of a force and one element Maximum acceleration. As show the image below:

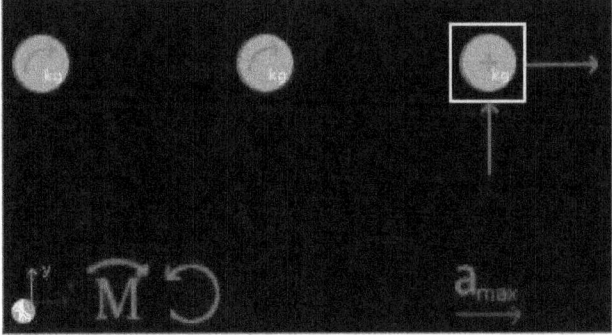

The element Stationary reference system write to default the gravity value. And the time is unknown.

g   9.81

t   time

The mass of the block is not provided, since it is irrelevant to what is asked. But as we need to specify a mass, will place 1 kg. Thus, for the element Mobile with circular movement placed at left, which represents the initial state, we have:

Name	initial
m	1
vt	0
r	40 @ cm
y	0

For the another element Mobile with circular movement, placed at right, which represents the final state, we have:

Name	final
m	1
vt	1.2
r	40 @ cm
y	0

Now we add both objects to element Angular momentum, where M is the moment that causes the acceleration of the block:

System i	initial
System f	final
M	M

The element Mobile with perpendicular circular motion is used to set additional conditions on the final state. Calling it A and leaving the center of rotation to the right, we have (both, the tangential acceleration and the tangential force, are unknowns):

Name	A
m	1
vt	1.2
r	40 @ cm
at	at

Ft   Ft

C   >

And in the element Maximum acceleration, we add the coefficient of friction, which ensures that both, the tangential force (Ft) and the centripetal force, are due to the maximum friction force:

Object   A

u   0.57

Now the element Moment of a force, is used to relate the maximum tangential force (Ft) with the applied moment:

M   M

f   Ft

d   40 @ cm

Finally, for normal and centripetal forces we have, respectively:

f   normal

f   centripetal

Now click in the icon Solve to get the answer:

```
centripetal = 3.600 N ;   time = 0.280 s ;   M = 1.711 N*m ;
at = 4.279 m/s2 ;   Ft = 4.279 N ;   normal = 9.810 N ;
Status = success.
```

The minimum time is 0.28 seconds.

13.3 Example 3

A 1.7 kg collar is attached to a spring and can sliding in a vertical plane (no friction) along the bar ABC. The spring constant is 550 N/m and is undeformed when its length is 16 cm. If the collar is released in A (V = 0), find its speed when passing through B.

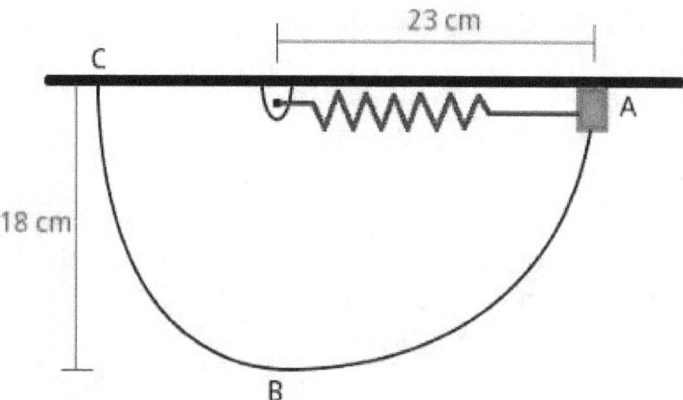

Solution with FisicaLab

Select the Dynamics group and, inside this, the Circular module. Erase the content of the chalkboard and select the SI system. And add one element Stationary reference system, one element Object in rest, one element Mobile with circular movement, two elements Spring, two elements Force, one element Initial system, one element Final system and one element Energy. As show the image below:

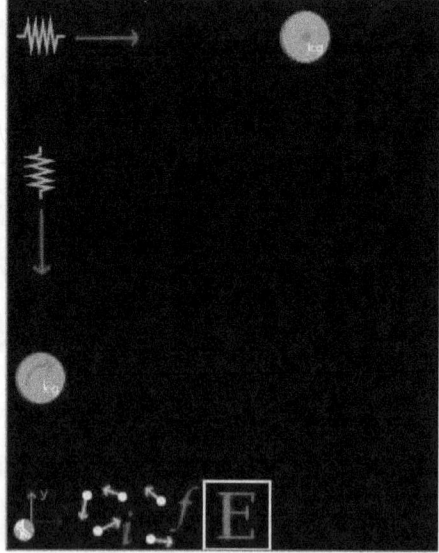

The element Stationary reference system write to default the gravity value. And the time is irrelevant for this problem. To the element Object in rest, which represents the collar in A, and taking this position as the horizontal reference:

Name	initialCollar
m	1.7
y	0

And for the element Spring corresponding to the initial state:

Name	initialSpring
k	550
x	7 @ cm

To the element Force applied to it:

f	initialForce

Both, the element Spring and the element Object in rest, corresponds to the initial state, so we add them to the element Initial system, that we call *initial*:

Name	initial
Object 1	initialCollar
Object 2	initialSpring
Object 3	0
Object 4	0

Now for the element Mobile with circular movement, representing the collar when passing through point B, we have (we do not know the data of the turning radius, but is irrelevant to this problem):

Name	finalCollar
m	1.7
vt	vt
r	0
y	-18 @ cm

For the element Spring that corresponds to the final state:

Name	finalSpring

k   550

x   2 @ cm

And to the applied force:

f   finalForce

Now add this element Spring and the element Mobile with circular movement, which represents the final state, to the element Final system, that we call *final*:

Name   final

Object 1 finalCollar

Object 2 finalSpring

Object 3 0

Object 4 0

Finally, add the elements initial and final system in the element Energy. And knowing that energy is conserved, we have:

System i initial

System f final

W   0

Now click in the icon Solve to get the answer:

```
vt = 2.233 m/s ;  initialForce = 38.500 N ;
finalForce = 11.000 N ;
Status = success.
```

13.4 Example 4

A 1.5 kg collar can slide without friction along a horizontal bar and is attached to two springs of constant 380 N/m, whose undeformed length is 2 cm. If the collar is released from rest in position A, find its speed when passing through point B.

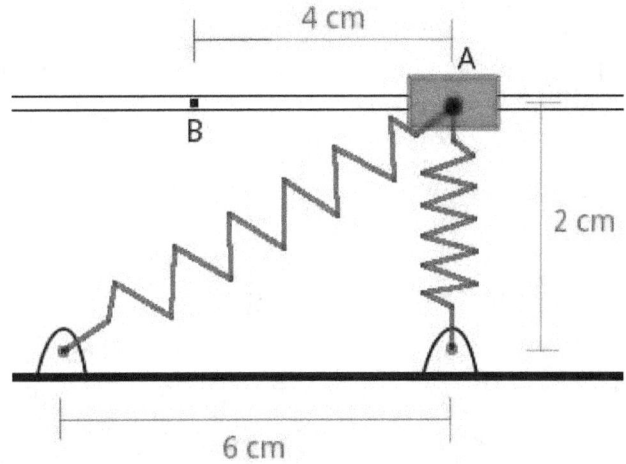

Solution with FisicaLab

Select the Dynamics group and, inside this, the Circular module. Erase the content of the chalkboard and select the SI system. And add one element Stationary reference system, one element Object in rest, one element Mobile with linear movement, four elements Spring, seven elements Force, one element Initial system, one element Final system and one element Energy. As show the image below:

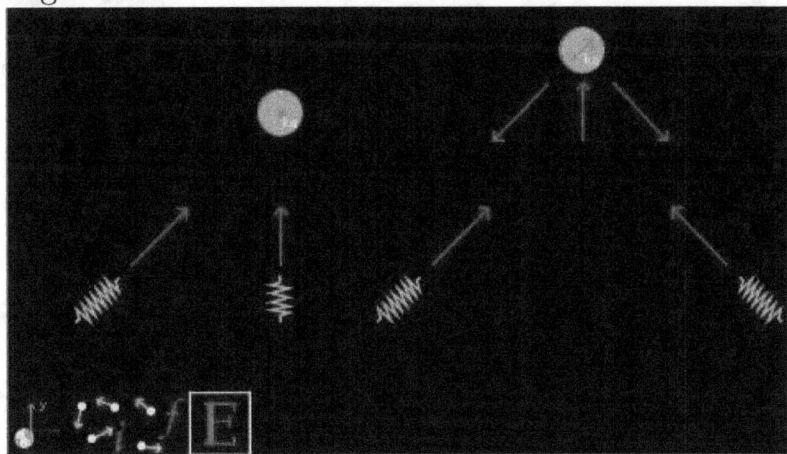

The element Stationary reference system write to default the gravity value. And the time is irrelevant for this problem. To the element Object in rest, which represents the collar in A, we have:

Name   initialCollar

m   1.5

y   0

To the Spring element in the vertical position, we have (this spring is neither stretched nor compressed):

Name   springOne

k   380

x   0

Now for the force applied to the spring:

f   forceOne

And to the second spring at the initial state, previously we get the stretched distance, we have (this operation can be done with `hypot(6,2)-2 cm`):

$$(\sqrt{6 \times 6 + 2 \times 2} - 2)\mathrm{cm} = 4.325\,\mathrm{cm}$$

Name   springTwo

k   380

x   4.325 @ cm

And to the applied force (here the angle is irrelevant):

f   forceTwo

ang   0

The element Object in rest and the two springs, are the initial state. Then we add them to the Initial system element, that we call *initial*:

Name   initial

Object 1   initialCollar

Object 2   springOne

Object 3   springTwo

Object 4   0

For the element Mobile with linear movement, which represents the collar at point B, and since the movement is horizontal (angle = 0) and the acceleration is unknown, we have (the coordinates x,y are irrelevant to this problem):

Name	finalCollar
m	1.5
v	v
x	0
y	0
ang	0
a	acceleration

For the vertical force, which represents the normal applied by the horizontal bar, we have:

f	normal

Now, to the force applied at left, the force applied by the spring to the left, and entering the angle as 2/2, we have:

f	forceThree
ang	45.000

And to the force applied at right, the force applied by the spring to the right, and entering the angle as 2/4, we have:

f	forceFour
ang	26.565

Now to the Spring element at left, previously we get the stretched distance (this operation can be done with `hypot(2,2)-2  cm`):

$$(\sqrt{2 \times 2 + 2 \times 2} - 2)\mathrm{cm} = 0.828\mathrm{cm}$$

Name	springThree
k	380
x	0.828 @ cm

And to the applied force (here the angle is irrelevant):

f   forceThree

ang   0

Now to the Spring element at right we have, previously we get the stretched distance (this operation can be done with `hypot(4,2)-2  cm`):

$$(\sqrt{4 \times 4 + 2 \times 2} - 2)\text{cm} = 2.472\text{cm}$$

Name   springFour

k   380

x   2.472 @ cm

And to the applied force (here the angle is irrelevant):

f   forceFour

ang   0

The element Mobile with linear movement and the two previous springs, are the final state. Then we add them to the Final system element, that we call *final*:

Name   final

Object 1   finalCollar

Object 2   springThree

Object 3   springFour

Object 4   0

Finally, add the elements initial and final system in the element Energy. And knowing that energy is conserved, we have:

System i   initial

System f   final

W   0

Now click in the icon Solve to get the answer:

```
normal = 21.141 N ;   forceThree = 3.146 N ;
forceFour = 9.394 N ;   forceTwo = 16.435 N ;
forceOne = -0.000 N ;   v = 0.549 m/s ;
acceleration = 4.118 m/s2 ;
Status = success.
```

The collar speed is 0,549 m/s.

13.5 Example 5

A collar of 430 grams can slide without friction along a horizontal rod, which in turn can rotate about a vertical axis. A spring of constant 4 N/m, whose undeformed length is 70 cm, holds the collar to the bar, as shown in the figure below. Initially the bar is turning with a constant angular velocity of 13 rad/s. While a rope holds the collar in the position A. Suddenly the rope breaks and the spring pushes the collar out. Neglecting the mass of the rod, find the tangential velocity of the collar to the position B and its radial acceleration.

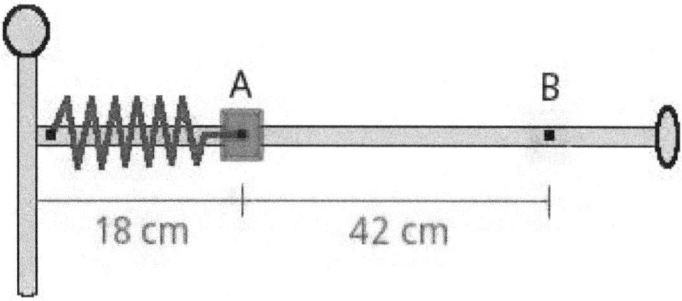

Solution with FisicaLab

Select the Dynamics group and, inside this, the Circular module. Erase the content of the chalkboard and select the SI system. And add one element Stationary reference system, one element Mobile with circular movement, one element Mobile with polar circular movement, two elements Spring, three elements Force, one element Initial system, one element Final system, one element Energy, one element Angular momentum and one element Angular velocity. As show the image below:

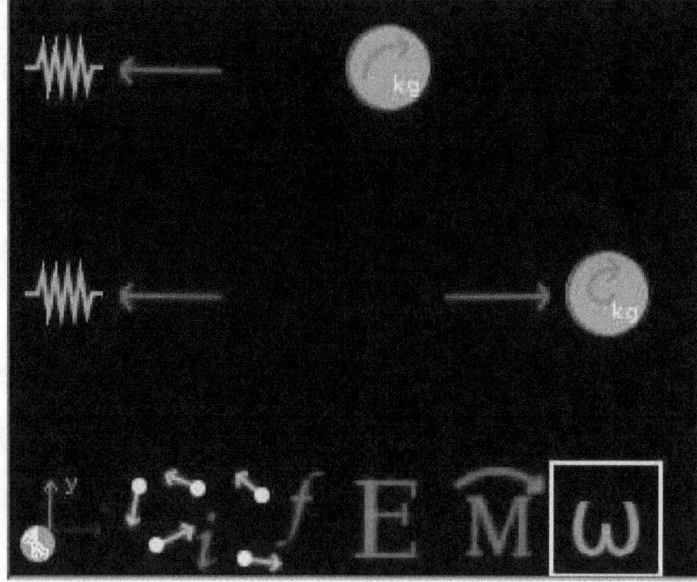

The element Stationary reference system write to default the gravity value. And the time is irrelevant for this problem. Now to the element

Mobile with circular movement, which represents the collar in A and taking in account that tangential velocity is unknown, we have:

Name	initialCollar
m	430 @ g
vt	vInitial
r	18 @ cm
y	0

And in the element Angular velocity we set the provided angular velocity for initial state:

Object	initialCollar
vang	13

For the Spring element of the initial state:

Name	initialSpring
k	4
x	-52 @ cm

And to the applied force:

f	initialForce

Now add this element Spring and the element Mobile with circular movement, which represents the initial state, to the element Initial system, that we call *initial*:

Name	initial
Object 1	initialCollar
Object 2	initialSpring
Object 3	0
Object 4	0

Now to the element Mobile with polar circular movement, which represents the collar in position B, we have:

Name	finalCollar

m	430 @ g
vt	vtFinal
r	60 @ cm
y	0
vr	vrFinal
ar	arFinal

And to the applied force to this element, the force applied by the spring, we have:

f	finalForce

This force is necessary to find the radial acceleration. Now to the element Spring of final state:

Name	finalSpring
k	4
x	-10 @ cm

And to the applied force:

f	finalForce

Now add this element Spring and the element Mobile with polar circular movement, which represents the final state, to the element Final system, that we call *final*:

Name	final
Object 1	finalCollar
Object 2	finalSpring
Object 3	0
Object 4	0

And add the elements initial and final system in the element Energy. And knowing that energy is conserved, we have:

System i	initial

System f   final

W   0

And in the element Angular momentum, we add the elements that represents the initial and final collar:

System i   initialCollar

System f   finalCollar

M   0

Now click in the icon Solve to get the answer:

```
initialForce = 2.080 N ;   finalForce = 0.400 N ;
vInitial = 2.340 m/s ;   vtFinal = 0.702 m/s ;
vrFinal = 2.721 m/s ;   arFinal = 0.930 m/s2 ;
Status = success.
```

13.6 Example 6

A system consisting of two small spheres of 900 and 350 grams, joined by a thin rod can rotate freely about a vertical axis passing through the center of mass of the system. Initially the system is at rest when a moment of 5 N*m is applied. Neglecting the mass of the rod, find the angular velocity and angular acceleration of the system after 1.2 seconds. Find also the power applied.

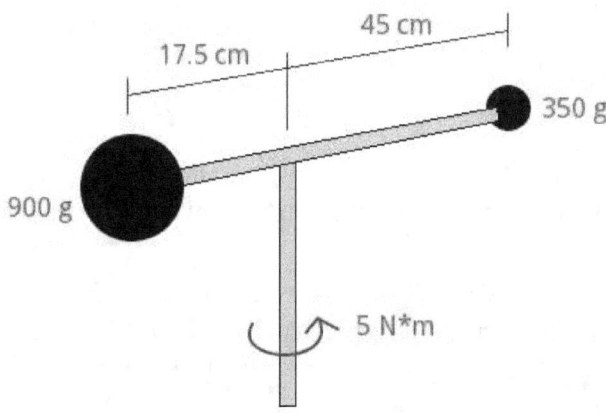

Solution with FisicaLab

Select the Dynamics group and, inside this, the Circular module. Erase the content of the chalkboard and select the SI system. And add one element Stationary reference system, four elements Mobile with circular movement, one element Mobile with perpendicular circular movement, two elements Force, one element Initial system, one element Final system, one element Power, one element Angular momentum, two elements Angular velocity, one element Angular acceleration, one element Moment of a force and one element Inertia. As show the image below:

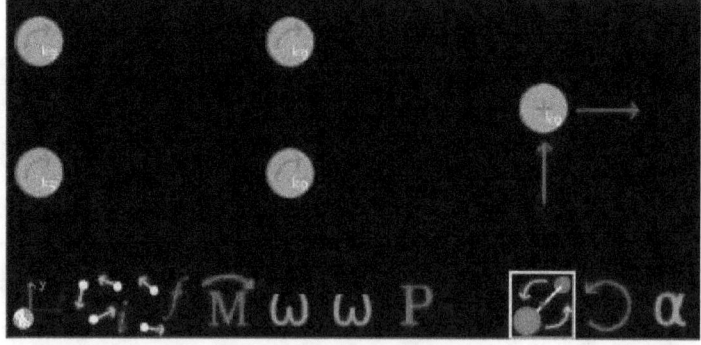

The element Stationary reference system write to default the gravity value. And the time is 1.2 seconds:

g   9.81

t   1.2

The two elements Mobile with circular movement at top represents the initial state of the system. Taking the left as the sphere of 900 grams, which we call *Ainitial*, and the right as the sphere of 350 grams, which we call *Binitial*, we have respectively:

Name	Ainitial
m	900 @ g
vt	0
r	17.5 @ cm
y	0

Name	Binitial
m	350 @ g
vt	0
r	45 @ cm
y	0

Now we add these to the Initial system element, which we call *initial*:

Name	initial
Object 1	Ainitial
Object 2	Binitial
Object 3	0
Object 4	0

The two elements Mobile with circular movement on the bottom represents the final state. Calling *Afinal* the sphere of 900 grams and *Bfinal* the sphere of 350 grams, we have:

Name	Afinal
m	900 @ g
vt	vtAfinal
r	17.5 @ cm
y	0

Name	Bfinal
m	350 @ g
vt	vtBfinal
r	45 @ cm
y	0

Now we add these to the Final system element, which we call *final*:

Name	final
Object 1	Afinal
Object 2	Bfinal
Object 3	0
Object 4	0

Now add the two systems to element Angular momentum, where we put also the applied momentum:

System i	initial
System f	final
M	5

Also add these to the Power element, where we put the power as an unknown:

System i	initial
System f	final
P	power

Now in Angular velocity elements, put the final angular velocity as an unknown (the same for both spheres):

Object	Afinal
vang	vang

Object	Bfinal

vang   vang

We use the element Mobile with perpendicular circular movement to know the angular acceleration, applying the force produced by the applied moment. But since the system have two spheres with different mass and different radius, first we get the total mass and the effective turning radius of the final system, using the Inertia element:

System	final
m	mTotal
r	radius

Now we can obtain the corresponding force for this system using the element Moment of a force:

M	5
f	Ft
d	radius

Now enter these information in the element Mobile with perpendicular circular movement, which we call *system*. As tangential velocity for this system we will place 0, the initial velocity (Since the angular acceleration is constant, has the same value at initial or final state. Anyway, we don't know the final velocity of this mobile.):

Name	system
m	mTotal
vt	0
r	radius
at	at
Ft	Ft
C	>

And to the applied forces:

f	normal

f	centripetal

Finally in the Angular acceleration element, we assign the Mobile with perpendicular circular movement:

Object   system

aang   aang

Now click in the icon Solve to get the answer:

```
normal = 12.263 N ;   centripetal = -0.000 N ;
mTotal = 1.250 kg ;   radius = 0.281 m ;
Ft = 17.817 N ;   aang = 50.794 rad/s2 ;
vtAfinal = 10.667 m/s ;   vtBfinal = 27.429 m/s ;
vang = 60.952 rad/s ;   power = 152.381 N*m/s ;
at = 14.254 m/s2 ;
Status = success.
```

13.7 Example 7

Two small spheres of 300 grams can slide without friction along a horizontal thin bar, which in turn can rotate about a vertical axis. Initially the bar is rotating at a constant angular speed of 9 rad/s, while a rope holds both spheres at a distance of 15 cm from the axis of rotation. Suddenly the rope breaks and the spheres stops at the ends of the bar. Find the final angular velocity and the loss energy. Also find the centripetal force and centripetal acceleration of the spheres at the beginning and end.

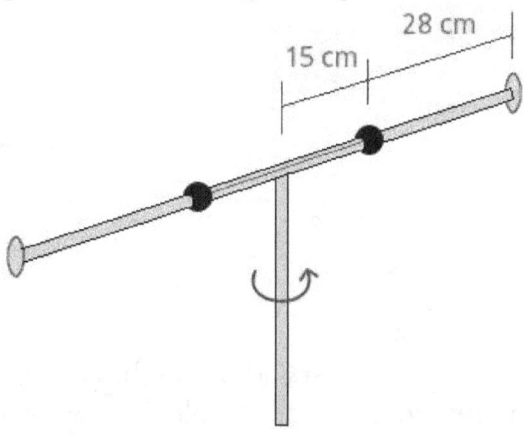

Solution with FisicaLab

Select the Dynamics group and, inside this, the Circular module. Erase the content of the chalkboard and select the SI system. And add one element Stationary reference system, four elements Mobile with circular movement, two elements Force, one element Initial system, one element Final system, one element Energy, one element Angular momentum, two elements Angular velocity and two elements Angular acceleration. As show the image below:

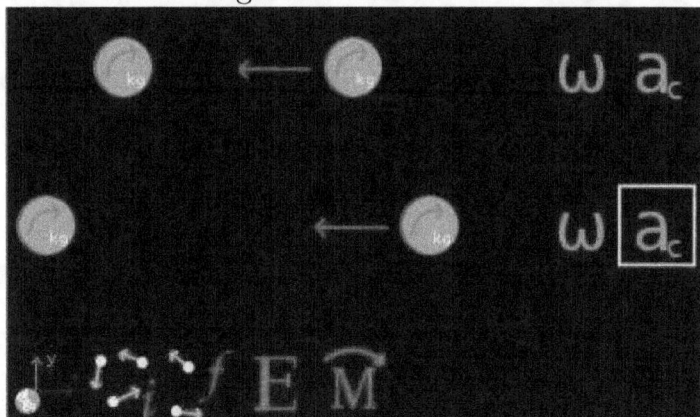

The element Stationary reference system write to default the gravity value. And the time is irrelevant for this problem. Now the two elements Mobile with circular movement at top represents the initial state of the system. Calling them *Ainitial* and *Binitial*, we have:

Name	Ainitial
m	300 @ g
vt	vtInitial
r	15 @ cm
y	0

Name	Binitial
m	300 @ g
vt	vtInitial
r	15 @ cm
y	0

Both have the same tangential speed at the beginning, which is unknown. Now with one of the elements Angular velocity, we set the initial angular velocity. You only need to do this for one of the spheres:

Object	Ainitial
vang	9

For the element Force applied to one of these spheres, that represents the centripetal force, we have:

f	centInitial

We only add one force because both spheres have the same centripetal force. Now for one of the Centripetal acceleration elements:

Object	Ainitial
ac	acInitial

Now we add these spheres to the Initial system element, which we call *initial*:

Name	initial
Object 1	Ainitial
Object 2	Binitial

Object 3   0

Object 4   0

Now the two elements Mobile with circular movement at bottom represents the final state of the system. Calling them *Afinal* and *Bfinal*, we have:

Name	Afinal
m	300 @ g
vt	vtFinal
r	43 @ cm
y	0

Name	Bfinal
m	300 @ g
vt	vtFinal
r	43 @ cm
y	0

Both have the same tangential speed at final. For the applied force:

f	centFinal

And to elements Angular velocity and Centripetal acceleration we have, respectively:

Object	Afinal
vang	vangFinal

Object	Afinal
ac	acFinal

Now we add these spheres to the Final system element, which we call *final*:

Name	final

Object 1 Afinal

Object 2 Bfinal

Object 3 0

Object 4 0

And add the elements initial and final system in the element Angular momentum, remember that there isn't an applied moment:

System i initial

System f final

M 0

Also add the elements initial and final system in the element Energy:

System i initial

System f final

W energy

Now click in the icon Solve to get the answer:

```
vangFinal = 1.095 rad/s ;  acInitial = 12.150 m/s2 ;
acFinal = 0.516 m/s2 ;  vtInitial = 1.350 m/s ;
vtFinal = 0.471 m/s ;  centInitial = 3.645 N ;
centFinal = 0.155 N ;  energy = -0.480 N*m ;
Status = success.
```

13.8 Example 8

A small disc of 500 grams and another of 1.5 kg are tied by a rope of 48 cm. Both disks rotate in the direction of clockwise with an angular velocity of 8 rad/s around the center of mass. At the time shown in the image, the center of mass intersects the X axis at 2 meters from the origin with an speed of 0.45 m/s at 55 degrees from the horizontal. Moments after the rope breaks, and it is observed that the disk of 1.5 kg is moved parallel to the axis X and the disk of 500 grams intersects the Y axis at a distance of 4.6 m from the origin. Find a) The speed of both discs after the rope breaks and b) the distance d between the X axis and the trajectory of the disk of 1.5 kg.

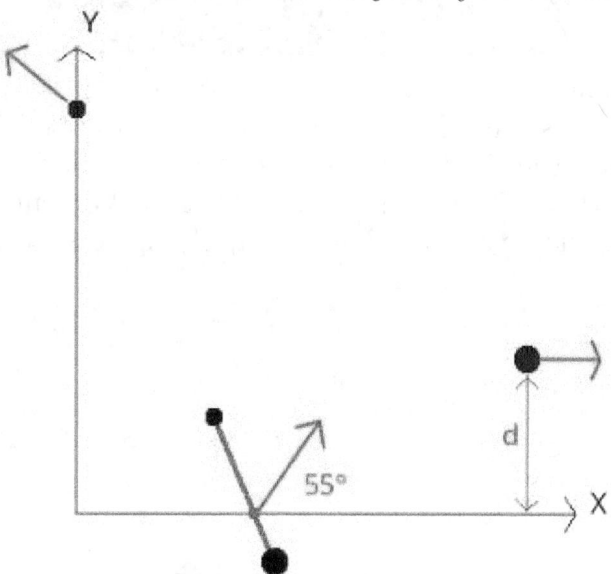

Solution with FisicaLab

Select the Dynamics group and, inside this, the Circular module. Erase the content of the chalkboard and select the SI system. And add two elements Mobile with circular movement, two elements Mobile with linear movement, one element Center of rotation, one element Initial system, one element Final system, two elements Angular velocity, one element Energy, one element Angular momentum and one element Linear momentum. As show the image below:

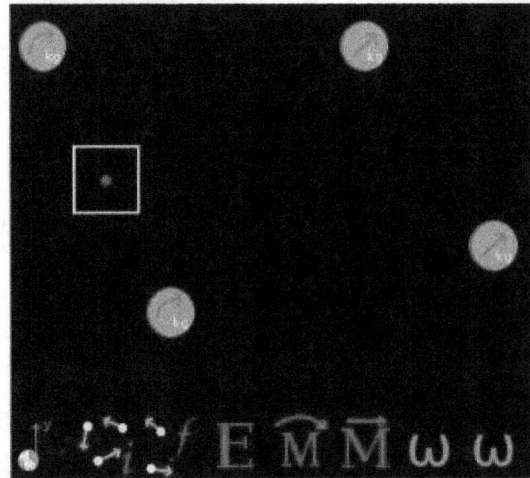

The two elements Mobile with circular movement at left represent the initial state of the system. Taking the left as the disk of 500 grams, which we call *Ainitial*, and the right as the disk of 1.5 kg, which we call *Binitial*, we have respectively (the distance from each disk to the center of rotation can be obtained using the `rd()` function and applying the conversion factor):

Name	Ainitial
m	500 @ g
vt	vtA
r	36 @ cm
y	0

Name	Binitial
m	1.5
vt	vtB
r	12 @ cm
y	0

The tangential velocity is unknown. But we know the angular velocity. Therefore to Angular velocity elements, remember that rotation is in sense of clockwise:

Object	Ainitial
vang	-8

Object	Binitial

vang	-8

Now for the element Center of rotation, which also forms part of the initial state:

Name	center
v	0.45
ang	55
x	2
y	0

As the two elements Mobile with circular movement and Center of rotation are the initial state, we add them to the element Initial system, which we call *initial*:

Name	initial
Object 1	centtr
Object 2	Ainitial
Object 3	Binitial
Object 4	0

Now for the element Mobile with linear movement at top, that represent the disk of 500 grams that intersect the axis Y, we have, calling it *Afinal*:

Name	Afinal
m	500 @ g
v	vAfinal
x	0
y	4.6
ang	angA
a	0

For the other element Mobile with linear movement at bottom, which represents the disk of 1.5 kg which travels parallel to the X axis at a distance d, and calling it *Bfinal*:

Name	Bfinal

m	1.5
v	vBfinal
x	0
y	d
ang	0
a	0

These two elements form the final state, then we add them to the element Final system, which we call *final*:

Name	final
Object 1	Afinal
Object 2	Bfinal
Object 3	0
Object 4	0

Now for the element Energy, knowing that energy is conserved, we have:

System i	initial
System f	final
W	0

To the element Angular momentum, taking into account that no external moments applied:

System i	initial
System f	final
M	0

And to element Linear momentum, no external forces are applied, we have:

System i	initial
System f	final
Fx	0
Fy	0

Now click in the icon Solve to get the answer:

```
vBfinal = 1.145 m/s ;   d = 2.761 m ;   vtA = -2.880 m/s ;
vtB = -0.960 m/s ;   vAfinal = 2.818 m/s ;
angA = 148.447 degrees ;
Status = success.
```

13.9 Example 9

A child of 30 kg is playing on an ice rink. The child moves at a linear speed of 5 m/s when passing beside a box of 9 kg, which has a rope tied. If the child takes the rope, and once stretched the separation between the child and the box is 6.5 meters, describe the subsequent move of child-box system.

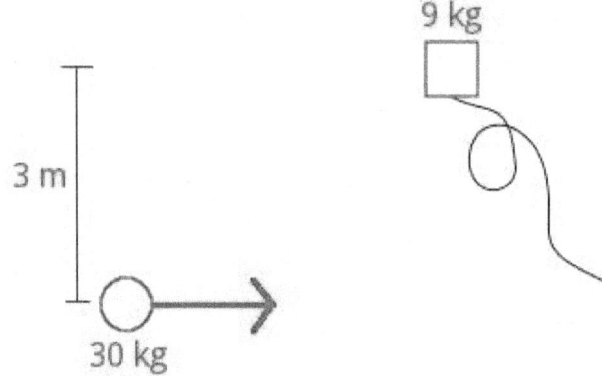

Solution with FisicaLab

Select the Dynamics group and, inside this, the Circular module. Erase the content of the chalkboard and select the SI system. And add one element Mobile with linear movement, two elements Mobile with circular movement, one element Center of rotation, one element Final system, two elements Angular velocity, one element Angular momentum and one element Linear momentum. As show the image below:

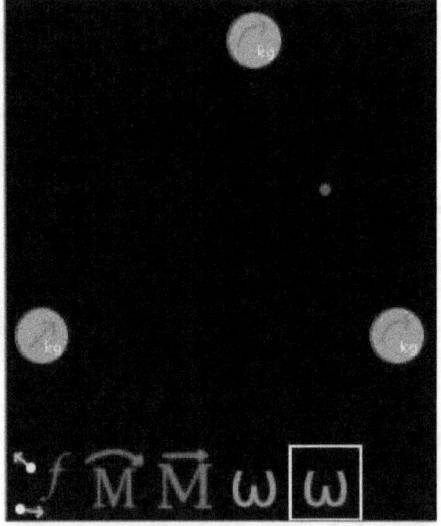

The element Mobile with linear movement represents the child in the initial state. Assuming initially that moves along the X axis, we have:

Name   initialChild

m   30

v	5
x	0
y	0
ang	0
a	0

Now the element Center of rotation represents the center around which the child and the box tour at the final state. Noting that the angle of inclination of the rope, once taut, with respect to the horizontal is 27.486 degrees, and the distance from the center of rotation to child is 1.5 meters, we can obtain the coordinates of the center of rotation as (these operations can be done directly at text fields):

```
x = 1.5*cos(27.486) = 1.331
y = 1.5*sin(27.486) = 0.692
```

Thus we have for this element:

Name	center
v	vCenter
ang	angCenter
x	1.331
y	0.692

Now the element Mobile with circular movement at top, represents the box, and calculating the distance from this to the center of rotation (5 m, the function **rd()** can be used to get this value), we have:

Name	finalBox
m	9
vt	vtBox
r	5
y	0

The other element Mobile with circular movement to the lower right side, represents the child in the final state, and calculating the distance of the center of rotation (1.5 m, the function **rd()** can be used to get this value), we have:

Name	finalBoy
m	30
vt	vtBoy
r	1.5
y	0

The box and the child have the same angular velocity. Therefore for the Angular velocity elements, respectively:

Object	finalBoy
vang	vang

Object	finalBox
vang	vang

The element Center of rotation and the two elements Mobile with circular movement are the final state. Then we add them to the element Final system, which we call *final*:

Name	final
Object 1	center
Object 2	finalBoy
Object 3	finalBox
Object 4	0

To the element Angular momentum, taking into account that no external moments applied:

System i	initialBoy
System f	final
M	0

And to element Linear momentum, no external forces are applied, we have:

System i	initialBoy

System f   final

Fx   0

Fy   0

Now click in the icon Solve to get the answer:

```
vang = 0.355 rad/s ;   vCenter = 3.846 m/s ;
angCenter = 360.000 degrees ;   vtBox = 1.774 m/s ;
vtBoy = 0.532 m/s ;
Status = success.
```

The center of rotation moves horizontally to the right with a speed of 3.846 m/s. And the boy and the box rotate in a counter-clockwise with an angular velocity of 0.355 rad/s.

Note: *Note that in this problem the mass of 9 kg does not appear in the initial state. This is because for these problems FisicaLab allows only one object in the initial state. The object that initially is moving.*

13.10 Example 10

Three masses are on an horizontal frictionless surface. Two of these, 3 and 6 kg, are at rest and are linked by a lightweight rod. The third, of 1 kg, moves with a speed of 4 m/s as shown in the image. If this collide with mass of 3 kg and bounces at 90 degrees with a speed of 2 m/s. What will be the subsequent movement of the other two masses? Also obtain the absolute velocity of the masses of 3 and 6 kg just after impact.

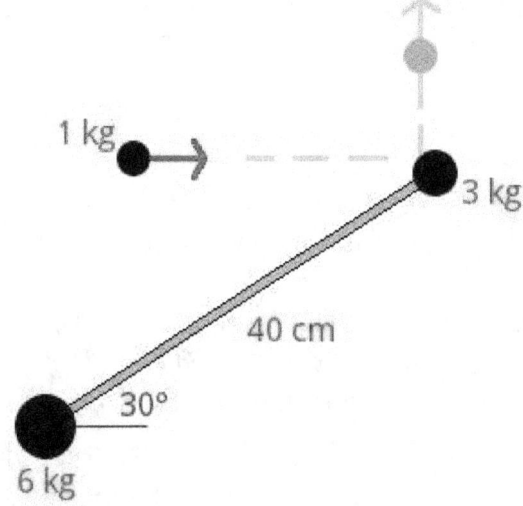

Solution with FisicaLab

Select the Dynamics group and, inside this, the Circular module. Erase the content of the chalkboard and select the SI system. And add two elements Mobile with linear movement, two elements Mobile with circular movement, one element Center of rotation, one element Final system, two elements Angular velocity, two elements Absolute velocity, one element Angular momentum and one element Linear momentum. As show the image below:

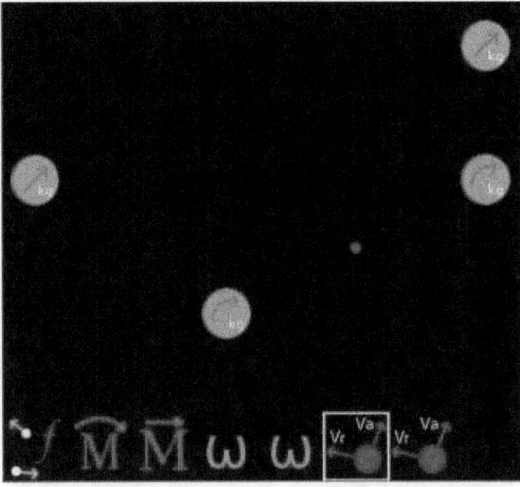

The element Mobile with linear movement to the left represents the mass of 1 kg in the initial state. Assuming the mass of 6 kg is at the origin of the coordinate system, we have for this element, called *initial*:

Name	initial
m	1
v	4
x	0
y	20 @ cm
ang	0
a	0

To the other Mobile with linear movement, which represents the same mass of 1 kg but in the final state, we have, calling it *A* (the X coordinate is `40*cos(30)`, this can be operated directly at text field):

Name	A
m	1
v	2
x	34.641 @ cm
y	0
ang	90
a	0

To the element Center of rotation, around which the masses of 3 and 6 kg rotate in the final state, we first find its coordinates. Noting that the distance from the mass of 6 kg is 13,333 m, we have (these operations can be done directly at text fields):

```
x = 13.333*cos(30) = 11.547
y = 13.333*sin(30) = 6.666
```

Thus we have for this element:

Name	center
v	vCenter
ang	angCenter

x   11.547 @ cm

y   6.666 @ cm

Now for the element Mobile with circular movement to the right, that represents the mass of 3 kg in the final state, we have, calling it B (the distance from the center can be calculated with the function **rd()**):

Name	B
m	3
vt	vtB
r	26.666 @ cm
y	0

To the other element Mobile with circular movement, that represents the mass of 6 kg in the final state, we have, calling it C (the distance from the center can be calculated with the function **rd()**):

Name	C
m	6
vt	vtC
r	13.333 @ cm
y	0

These two elements have the same angular velocity. Therefore for the Angular velocity elements, respectively:

Object	B
vang	vang

Object	C
vang	vang

These three mobiles and the center of rotation constitute the final state, then we add them to the element Final system, that we call *final*:

Name	final

Object 1   center

Object 2   A

Object 3   B

Object 4   C

Since are requested the absolute velocity of masses 3 and 6 kg (B and C), we have for the elements Absolute velocity, respectively:

Object	B
angR	30
Center	center
v	vB
ang	angB

Object	C
angR	210
Center	center
v	vC
ang	angC

To the element Angular momentum, taking into account that no external moments applied:

System i	initial
System f	final
M	0

And to element Linear momentum, no external forces are applied, we have:

System i	initial
System f	final
Fx	0
Fy	0

Now click in the icon Solve to get the answer:

```
vang = -3.110 rad/s ;   vtC = -0.415 m/s ;
vB = 1.274 m/s ;   angB = 312.412 degrees ;
vtB = -0.829 m/s ;   vC = 0.274 m/s ;
angC = 30.004 degrees ;   vCenter = 0.497 m/s ;
angCenter = 333.435 degrees ;
Status = success.
```

Note: *Note that in this problem the masses of 3 and 6 kg does not appear in the initial state. This is because for these problems FisicaLab allows only one object in the initial state. The object that initially is moving. See previous example.*

13.11 Example 11

A mass of 0.1 slugs is released from rest in A and slides without friction along the track shown. If the radius of curvature of the bottom is 3 ft, find the force applied by the track to the mass when it passes through the point B.

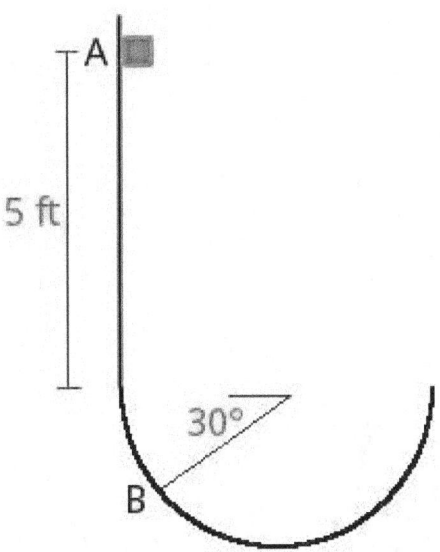

Solution with FisicaLab

Select the Dynamics group and, inside this, the Circular module. Erase the content of the chalkboard and select the English system. And add one element Stationary reference system, one element Object in rest, one element Mobile with circular movement, one element Force and one element Energy. As show the image below:

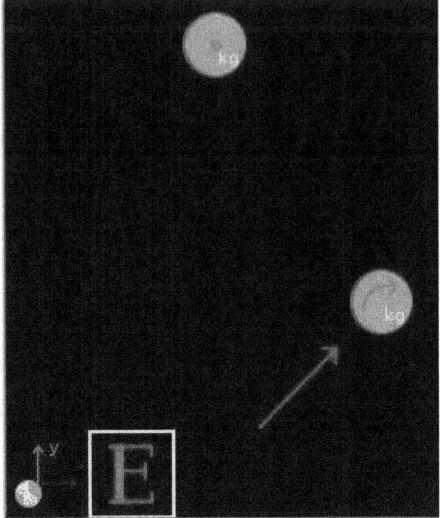

The element Stationary reference system write to default the gravity value. And the time is irrelevant for this problem. Now the element Object in rest represents the mass at point A. If we take as horizontal reference the position B, the Y coordinate of the mass shall be: `5 ft + 3*sin(30) ft = 6.5 ft` (this can be operated directly at text field). Then we have for this element:

Name	A
m	0.1
y	6.5

Now to the element Mobile with circular movement, which represents the mass at point B, we have (the tangential velocity is unknown):

Name	B
m	0.1
vt	vt
r	3
y	0

And to the element Force, which represents the normal force applied by the track:

f	normal
ang	30

Now add the elements initial and final system in the element Energy. And knowing that energy is conserved, we have:

System i	A
System f	B
W	0

Now click in the icon Solve to get the answer:

```
normal = 15.563 lb ;   vt = 20.460 ft/s ;
Status = success.
```

13.12 Example 12

A mass of 3 kg traveling at 32 m/s, as shown in the figure below, explodes separating into two fragments. If these pass through the coordinates shown and the 1.8 kg fragment has the shown direction, find the speed of both fragments, as well as the direction of the fragment of 1.2 kg.

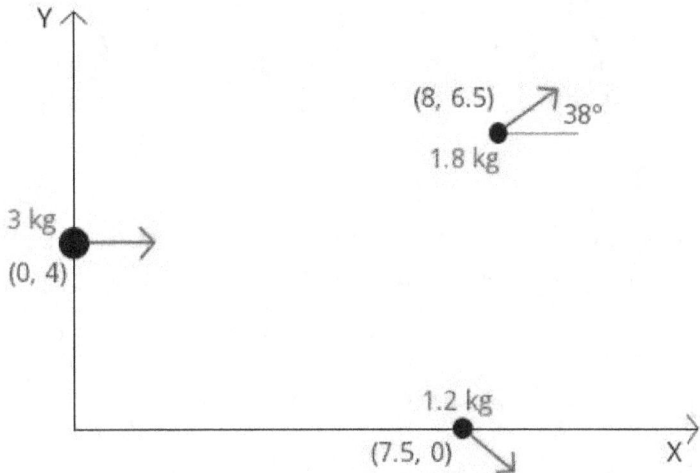

Solution with FisicaLab

Select the Dynamics group and, inside this, the Circular module. Erase the content of the chalkboard and select the SI system. And add three elements Mobile with linear movement, one element Final system, one element Angular momentum and one element Linear momentum. As show the image below:

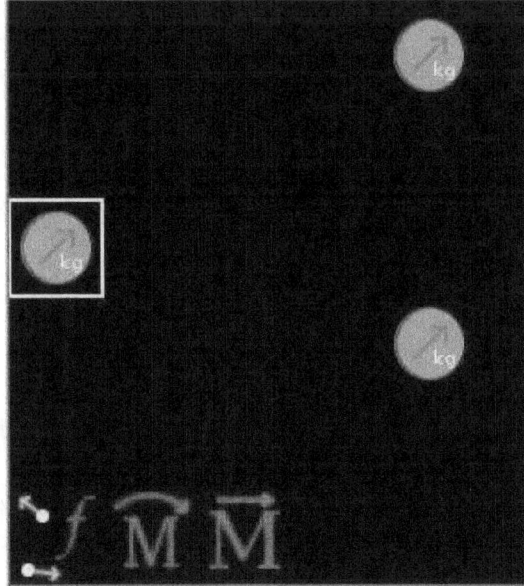

To the element Mobile with linear movement to the left, that represents the mass of 3 kg in the initial state, we have, calling it A:

Name	A
m	3
v	32
x	0
y	4
ang	0
a	0

Now for the element Mobile with linear movement at top, that represents the fragment of 1.8 kg, we have, calling it B:

Name	B
m	1.8
v	vB
x	8
y	6.5
ang	38
a	0

And to the element Mobile with linear movement at bottom, that represents the fragment of 1.2 kg, we have, calling it C:

Name	C
m	1.2
v	vC
x	7.5
y	0
ang	angC
a	0

The fragments B and C are the final state, so we add them to the element Final system, which we call *final*:

Name	final
Object 1	B

Object 2   C

Object 3   0

Object 4   0

Now to the element Angular momentum, taking into account that no external moments applied:

System i   A

System f   final

M   0

And to element Linear momentum, no external forces are applied, we have:

System i   A

System f   final

Fx   0

Fy   0

Now click in the icon Solve to get the answer:

```
vB = 44.313 m/s ;   vC = 49.372 m/s ;
angC = 304.018 degrees ;
Status = success.
```

13.13 Example 13

A small sphere of 400 grams, tied to a rope of 60 cm, can oscillate in a vertical plane. If the sphere is released from rest when the rope is stretched horizontally. And if the maximum tension that this support is 7 N, find the angle when the rope breaks.

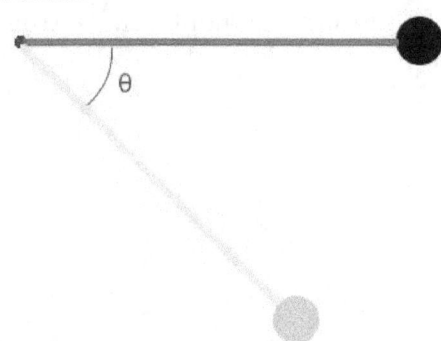

Solution with FisicaLab

Select the Dynamics group and, inside this, the Circular module. Erase the content of the chalkboard and select the SI system. And add one element Stationary reference system, one element Object in rest, one element Mobile with circular movement, one element Force, one element Energy and one element Sine of angle. As show the image below:

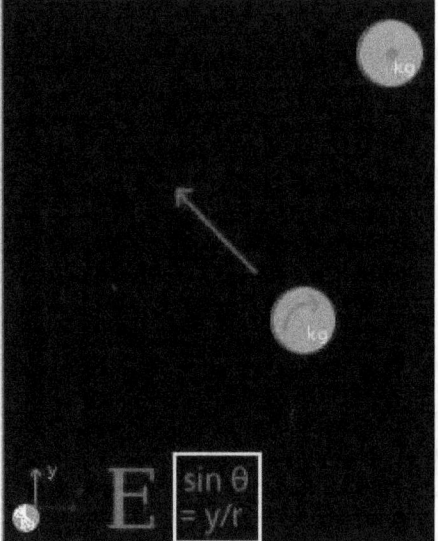

The element Stationary reference system write to default the gravity value. And the time is irrelevant for this problem. Now the element Object in rest represents the initial state. If we take as horizontal reference this position, we have for this element:

Name   A

m   400 @ g

y   0

To the element Mobile with circular movement, which represents the final state, and taking into account that both the speed and height are unknowns, we have:

Name	B
m	400 @ g
vt	vt
r	60 @ cm
y	yB

And to the Force element, which represents the force applied by the rope and taking into account that the angle of the force is the same as that of the rope, we have:

f	7
ang	ang

Now for the Sine of angle element, where we relate the height of the sphere in the final state and the turning radius with the unknown angle, we have:

y	yB
r	60 @ cm
ang	ang

Finally for the element Energy, and knowing that energy is conserved, we have:

System i	A
System f	B
W	0

Now click in the icon Solve to get the answer:

```
yB = -0.357 m ;   ang = 36.486 degrees ;   vt = 2.646 m/s ;
Status = success.
```

14 Calorimetry

In this module only the SI system is available, with the following conversion factors:

T	metric ton
g	gram
slug	slug
C	degree Celsius
R	Rankin
F	degree Farenheit
cm	centimeter
mm	millimeter
ft	feet
in	inch
h	hour
min	minute
cm2	squared centimeters
mm2	squared millimeters
ft2	squared feet
in2	squared inch
cm3	cubic centimeters
mm3	cubic millimeters
ft3	cubic feet
in3	cubic inch
L	liter
psi	pound per squared inch
atm	atmosphere
bar	bar
Torr	Torricelli
mmHg	millimeters of mercury
J/kg*C	Joules per kilogram Celsius

cal/g*K	calorie per gram Kelvin
kcal/kg*K	kilocalorie per kilogram Kelvin
cal/g*C	calorie per gram Celsius
kcal/kg*C	kilocalorie per kilogram Celsius
cal/g	calorie per gram
kcal/kg	kilocalorie per kilogram
kWh	kilo Watts hour
Btu	British thermal unit
kcal	kilocalorie
cal	calorie
hp	horsepower
Btu/h	British thermal unit per hour
lb*ft/s	pound feet per second
m3/min	cubic meters per minute
ft3/s	cubic feet per second
gal/min	gallons per minute
L/min	liters per minute

This module have 20 elements, described below. With a description of each one, its data and its number of equations. In parentheses are shown the units used be FisicaLab (if no conversion factors are used) in cases where there may be doubt. The elements Block, Liquid, Gas, Change of state solid-liquid, Change of state liquid-gas and Process, support the application of heat elements (Heat applied, Applied heat flow, Heat extracted and Refrigeration). The heat elements are applied to other element placing these in one of the eight cells around.

14.1 Laboratory clock

The clock to measure the time.

Equations: None.

Data:
t: Time during which supplied (or extracted)
                a heat flow.

14.2 Applied heat

A certain amount of heat applied.

Equations: None.

Data:
Q: Amount of heat applied (in Joules).

14.3 Applied heat flow

Applied heat flow.

Equations: None.

Data:
dQ/dt: Applied heat flow (in Joules per second,
                Watts).

14.4 Heat extracted

A certain amount of heat extracted.

Equations: None.

Data:
Q: Amount of heat extracted (in Joules).

14.5 Refrigeration

Extracted heat flow.

Equations: None.

Data:
dQ/dt: Extracted heat flow (in Joules per second, Watts).

14.6 Block

Some amount of matter in solid state.

Equations: 1

Data:

Name: Name of the block.

m: Mass.

c: Specific heat of material (in J/kg*K).

Ti: Initial temperature (in Kelvins).

Tf: Final temperature (in Kelvins).

14.7 Liquid

Some amount of matter in liquid state.

Equations: 1

Data:

Name: Name of the liquid.

m: Mass.

c: Specific heat of liquid (in J/kg*K).

Ti: Initial temperature (in Kelvins).

Tf: Final temperature (in Kelvins).

14.8 Gas

Some amount of matter in gas state.

Equations: 1

Data:

Name: Name of the gas.

m: Mass.

c: Specific heat of gas (in J/kg*K).

Ti: Initial temperature (in Kelvins).

Tf: Final temperature (in Kelvins).

14.9 Linear expansion

Bar that performs a linear expansion (or contraction).

Equations: 1

Data:

k: Coefficient of linear expansion (in 1/K).

Li: Initial length.

Lf: Final length.

Ti: Initial temperature (in Kelvins).

Tf: Final temperature (in Kelvins).

14.10 Superficial expansion

Surface that performs a superficial expansion (or contraction).

Equations: 1

Data:

k: Coefficient of surface expansion (in 1/K).

Si: Initial surface.

Sf: Final surface.

Ti: Initial temperature (in Kelvins).

Tf: Final temperature (in Kelvins).

14.11 Volumetric expansion

Volume that performs a volumetric expansion (or contraction).

Equations: 1

Data:

k: Coefficient of volumetric expansion (in 1/K).

Vi: Initial volume.

Vf: Final volume.

Ti: Initial temperature (in Kelvins).

Tf: Final temperature (in Kelvins).

14.12 Change of state solid-liquid

A certain amount of matter in a change of state solid-liquid or vice versa.

Equations: 1

Data:

Name: Name of the change solid-liquid.

m: Mass of matter involved in the change.

cf: Heat of fusion (in Joules per kilogram).

Sense: Sense of change of state, > (greater than) indicates a change of state from solid to liquid. Whereas that < (less than) indicates the opposite process.

14.13 Change of state liquid-gas

A certain amount of matter in a change of state liquid-gas or vice versa.

Equations: 1

Data:

Name: Name of the change liquid-gas.

m: Mass of matter involved in the change.

cv: Heat of vaporization (in Joules per kilo-gram).

Sense: Sense of change of state, > (greater than) indicates a change of state from liquid to gas. Whereas that < (less than) indicates the opposite process.

14.14 Process

A certain amount of mass involved in a process with changes of state. The elements that may be involved in the process (5 in total, called Object 1, Object 2,...), are the blocks, liquids, gases and changes of state. Not allowed more than two changes of state per process. If the name of an object is left to '0', is assumed that this object is not involved. Also, the elements involved can not have applied heat elements (Heat applied, Applied heat flow, Heat extracted and Refrigeration).

Equations: 1

Data:

Name: Name of the process.

Object 1:   A block, liquid, gas or change of state involved in the process.

Object 2:   A block, liquid, gas or change of state involved in the process.

Object 3:   A block, liquid, gas or change of state involved in the process.

Object 4:   A block, liquid, gas or change of state involved in the process.

Object 5:   A block, liquid, gas or change of state involved in the process.

14.15 Calorimeter

Calorimeter for mixing. It can accommodate four processes, blocks, liquids or gases (called Object 1, Object 2 ,...), and does not allow elements Change of state. If the name of an object is left to '0', it is assumed that the object does not participate. Moreover, the elements involved (or elements of the processes involved) can not have applied elements heat (Heat applied, Applied heat flow, Heat extracted or Refrigeration).

Equations: 1

Data:

Object 1:   A block, liquid, gas or process involved in the mixing.

Object 2:   A block, liquid, gas or process involved in the mixing.

Object 3:   A block, liquid, gas or process involved in the mixing.

Object 4:   A block, liquid, gas or process involved in the mixing.

14.16 Gas at constant pressure

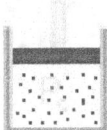

An ideal gas at constant pressure (law of Charles and Gay-Lussac).

Equations: 1

Data:

Vi: Initial volume.

Ti: Initial temperature (in Kelvins).

Vf: Final volume.

Tf: Final temperature (in Kelvins).

14.17 Gas at constant temperature

An ideal gas at constant temperature (law of Boyle-Mariotte).

Equations: 1

Data:

Pi: Initial pressure (in Pascals).

Vi: Initial volume.

Pf: Final pressure (in Pascals).

Vf: Final volume.

14.18 Gas at constant volume

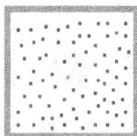

An ideal gas at constant volume.

Equations: 1

Data:

Pi: Initial pressure (in Pascals).

Ti: Initial temperature (in Kelvins).

Pf: Final pressure (in Pascals).

Tf: Final temperature (in Kelvins).

14.19 Gas PV/T

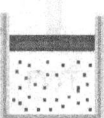

An ideal gas.

Equations: 1

Data:

Pi: Initial pressure (in Pascals).

Vi: Initial volume.

Ti: Initial temperature (in Kelvins).

Pf: Final pressure (in Pascals).

Vf: Final volume.

Tf: Final temperature (in Kelvins).

14.20 Heat exchanger

A simple heat exchanger.

Equations: 1

Data:

TRi: Inlet temperature of refrigerant (in Kelvins).

TRf: Outlet temperature of refrigerant (in Kelvins).

dR/dt: Flow of refrigerant (in cubic meters per second).

cR: Specific heat of refrigerant (in J/kg*K).

TFi: Initial temperature of fluid (in Kelvins).

TFf: Outlet temperature of fluid (in Kelvins).

dF/dt: Flow of fluid (in cubic meters per second).

cF: Specific heat of fluid (in J/kg*K).

15 Examples calorimetry

15.1 Example 1

A block of copper (specific heat 390 J/kg*K) of 0.5 kg have an initial temperature of 17 C. If this block get 150 calories, what will be its final temperature in Celsius?

Solution with FisicaLab

Select the Thermodynamics group and, inside this, the Calorimetry and gases module. Erase the content of the chalkboard. And add one element Block and one element Applied heat. As show the image below:

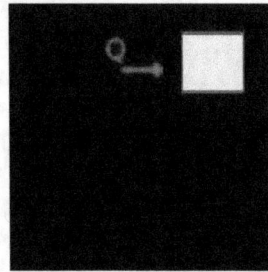

To the element Block, that represent the block of copper, we have:

Name	0
m	0.5
c	390
Ti	17 @ C
Tf	tf @ C

And to the element Applied heat:

Q	150 @ cal

Now click in the icon Solve to get the answer:

```
tf = 20.221 C ;
Status = success.
```

15.2 Example 2

An electric heater of 1500 Watts is placed in 10 kilograms of water (specific heat 4187 J/kg*K) with an initial temperature of 24 C. What will be the final temperature of the water in Celsius, after 8 minutes (assuming that all the heat generated by the heater is absorbed by water)?

Solution with FisicaLab

Select the Thermodynamics group and, inside this, the Calorimetry and gases module. Erase the content of the chalkboard. And add one element Liquid, one element Applied heat flow and one element Laboratory clock. As show the image below:

To the element Liquid, the 10 kg of water, we have:

Name	0
m	10
c	4187
Ti	24 @ C
Tf	tf @ C

Now to the element Applied heat flow:

dQ/dt	1500

And to the element Laboratory clock:

t	8 @ min

Now click in the icon Solve to get the answer:

```
tf = 41.196 C ;
Status = success.
```

15.3 Example 3

A bar of aluminum (coefficient of linear expansion 24E-6 1/K) of 55 cm and an initial temperature of 20 C, increases its temperature to 100 C. What is the length of the bar at this temperature?

Solution with FisicaLab

Select the Thermodynamics group and, inside this, the Calorimetry and gases module. Erase the content of the chalkboard. And add one element Linear expansion as show the image below:

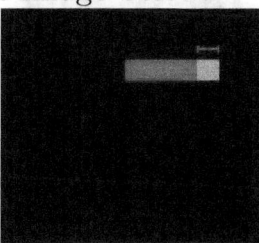

To this element, we have:

k	24E-6
Li	55 @ cm
Lf	Lf @ cm
Ti	20 @ cm
Tf	100 @ cm

Now click in the icon Solve to get the answer:

```
Lf = 55.106 cm ;
Status = success.
```

15.4 Example 4

At 24 C the volume of a recipient of copper is 1 Liter (coefficient of volumetric expansion 51E-6 1/K). What will be its volume, in liters, at 110 C?

Solution with FisicaLab

Select the Thermodynamics group and, inside this, the Calorimetry and gases module. Erase the content of the chalkboard. And add one element Volumetric expansion as show the image below:

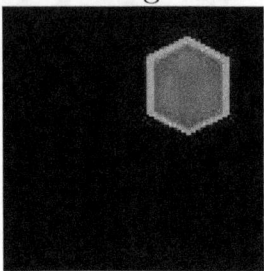

To this element, we have:

k	51E-6
Vi	1 @ L
Vf	Vf L
Ti	24 @ C
Tf	110 @ C

Now click in the icon Solve to get the answer:

```
Vf = 1.004 L ;
Status = success.
```

15.5 Example 5

How much heat must be removed to solidify 250 grams of cast iron at its melting point? Give your answer in Joules (heat of fusion of iron = 69.1 cal / g).

Solution with FisicaLab

Select the Thermodynamics group and, inside this, the Calorimetry and gases module. Erase the content of the chalkboard. And add one element Change of state solid-liquid and one element Heat extracted. As show the image below:

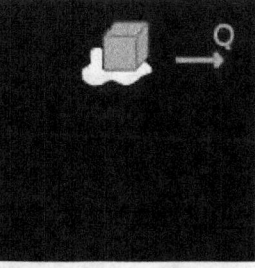

To the element Change of state, we have:

Name	0
m	250 @ g
cf	69.1 @ cal/g
Sense	<

In the entry of *Sense* we type < because the change of state is from liquid to solid. Now to the element Heat extracted:

Q	Q

Then click in the icon Solve to get the answer:

```
Q = 72330.425 J ;
Status = success.
```

15.6 Example 6

How much heat in calories must be added to transform a block of 3 kg of ice at -5 C, in water at + 22 C? (specific heat of ice = 0.50 cal/g*K, specific heat of water = 1 cal/g*K, and heat of fusion of water = 79.7 cal/g).

Solution with FisicaLab

Select the Thermodynamics group and, inside this, the Calorimetry and gases module. Erase the content of the chalkboard. And add one element Process, one element Applied heat (applied to the element Process), one element Block, one element Change of state solid-liquid and one element Liquid. As show the image below (the yellow arrows are only to indicate the direction of the process):

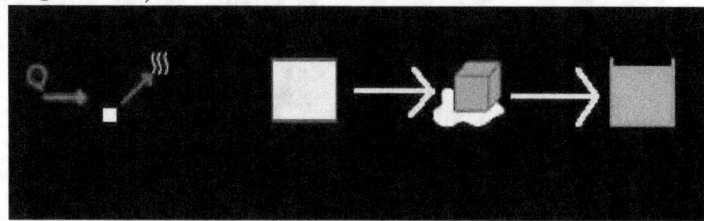

To the element Applied heat, with the conversion factor to kilocalories, we have:

Q   Q @ cal

The element Process, whose name in this issue is irrelevant, represents the change of the water from its solid state at -5 C until liquid state at 22 C. This element must contain the element Block, which represents the water in solid state, the element Change of state solid-liquid, which represents the change of state of the water, and the element Liquid, which represents the final state of the water. These elements gonna be called *Solid*, *Fusion* and *Liquid*, respectively. Then to the element Process we have:

Name   0

Object 1   Solid

Object 2   Fusion

Object 3   Liquid

Object 4   0

Object 5   0

Now to the element Block, which represents the water in solid state and that we will call *Solid*, because this name have in the element Process, we have:

Name	Solid
m	3
c	0.5 @ cal/g*K
Ti	-5 @ C
Tf	0 @ C

Now the element Change of state, called *Fusion*, is:

Name	Fusion
m	3
cf	79.7 @ cal/g
Sense	>

And to the element Liquid, which represents the final state of the water and called *Liquid*, we have:

Name	Liquid
m	3
c	1 @ cal/g*K
Ti	0 @ C
Tf	22 @ C

Now click in the icon Solve to get the answer:

```
Q = 312.600 kcal ;
Status = success.
```

15.7 Example 7

How much heat, in calories, should be removed to convert 25 g of molten lead at its melting point, in solid lead at 310 C? (Specific heat of the lead = 0.031 cal/g*K, heat of fusion of the lead = 5.54 cal/g, and melting temperature of the lead = 327 C).

Solution with FisicaLab

Select the Thermodynamics group and, inside this, the Calorimetry and gases module. Erase the content of the chalkboard. And add one element Process, one element Heat extracted (applied to the element Process), one element Change of state solid-liquid and one element Block. As show the image below (the yellow arrow are only to indicate the direction of the process):

To the element Heat extracted, we have, using the conversion factor to calories:

Q   Q @ cal

To the element Process (the elements Block and Change of state gonna be called *Solid* and *Fusion* respectively), we have:

Name	0
Object 1	Fusion
Object 2	Solid
Object 3	0
Object 4	0
Object 5	0

To the element Block, called *Solid*, we have:

Name	Solid
m	25 @ g
c	0.031 @ cal/g*K

Ti   327 @ C

Tf   310 @ C

Note that the initial temperature is 327 C and the final 310 C. Now to the element Change of state solid-liquid, called *Fusion*, we have care with direction of the change of state (sign <). Then:

Name   Fusion

m   25 @ g

cf   5.54 @ cal/g

Sense   <

Now click in the icon Solve to get the answer:

```
Q = 151.675 cal ;
Status = success.
```

15.8 Example 8

A rubber balloon inflated have a volume of 2000 cm3 at 18 C. If its temperature is increased at 30 C, What will be the new volume of the balloon (in cm3)?

Solution with FisicaLab

Select the Thermodynamics group and, inside this, the Calorimetry and gases module. Erase the content of the chalkboard. And add one element Gas at constant pressure, as show the image below:

To this element, with the corresponding conversion factors, we have:

Vi   2000 @ cm3

Ti   18 @ C

Vf   Vf @ cm3

Tf   30 @ C

Now click in the icon Solve to get the answer:

```
Vf = 2082.435 cm3 ;
Status = success.
```

15.9 Example 9

A certain gas have a volume of 5 liters, at 10 C and 750 mmHg of pressure. What will be the volume at 100 C and 3 atmospheres of pressure?

Solution with FisicaLab

Select the Thermodynamics group and, inside this, the Calorimetry and gases module. Erase the content of the chalkboard. And add one element Gas PV/T, as show the image below:

To this element, with the corresponding conversion factors, we have:

Pi   750 @ mmHg

Vi   5 @ L

Ti   10 @ C

Pf   3 @ atm

Vf   Vf @ L

Tf   100 @ C

Now click in the icon Solve to get the answer:

```
Vf = 2.162 L ;
Status = success.
```

15.10 Example 10

In a plastic cup of minimum heat capacity, we mix 300 g of water at 10 C, with 185 g of water at 100 C. What is the final temperature in Celsius? (Specific heat of water = 4187 J/kg*K).

Solution with FisicaLab

Select the Thermodynamics group and, inside this, the Calorimetry and gases module. Erase the content of the chalkboard. And add one element Calorimetry and two elements Liquid. As show the image below:

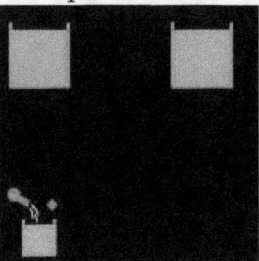

To one of the elements Liquid, we call this A and represent the 300 g of water, we have:

Name	A
m	300 @ g
c	4187
Ti	10 @ C
Tf	Tf @ C

To the other element Liquid, we call this B and represent the 185 g of water, we have:

Name	B
m	185 @ g
c	4187
Ti	100 @ C
Tf	Tf @ C

Note that the final temperature is the same unknown data in both elements, as must be. Now to the element Calorimetry:

Object 1   A

Object 2   B

Object 3   0

Object 4   0

Now click in the icon Solve to get the answer:

```
Tf = 44.330 C ;
Status = success.
```

15.11 Example 11

In a calorimeter with 325 g of water at 20 C, is introduced 77 g of an unknown metal at 90 C. If the final temperature is 26 C, what is the specific heat of the metal? (Specific heat of water = 4187 J/kg*K).

Solution with FisicaLab

Select the Thermodynamics group and, inside this, the Calorimetry and gases module. Erase the content of the chalkboard. And add one element Calorimetry, one element Block and one element Liquid. As show the image below:

To the element Liquid, we call this *Water*, we have:

Name	Water
m	325 @ g
c	4187
Ti	20 @ C
Tf	26 @ C

And to the element Block, we call this *Metal*, we have:

Name	Metal
m	77 @ g
c	cM
Ti	90 @ C
Tf	26 @ C

Note that the final temperature is the same in both elements, as must be. To the element Calorimetry, we have:

Object 1	Water
Object 2	Metal

Object 3   0

Object 4   0

Now click in the icon Solve to get the answer:

```
cM = 1656.788 J/kg*K ;
Status = success.
```

15.12 Example 12

A calorimeter of copper (105 g) contain 307 g of water at 23 C. If we add 95 g of ice at -4 C, What is the final temperature of the system in Celsius? (Specific heat of ice = 2090 J/kg*K, specific heat of water = 4187 J/kg*K, specific heat of copper = 390 J/kg*K, and heat of fusion to the water = 79.7 cal/g).

Solution with FisicaLab

Select the Thermodynamics group and, inside this, the Calorimetry and gases module. Erase the content of the chalkboard. And add one element Calorimetry, two elements Block, two elements Liquid, one element Change of state solid-liquid and one element Process. As show the image below (yellow circle contains the elements that represent the change of the ice):

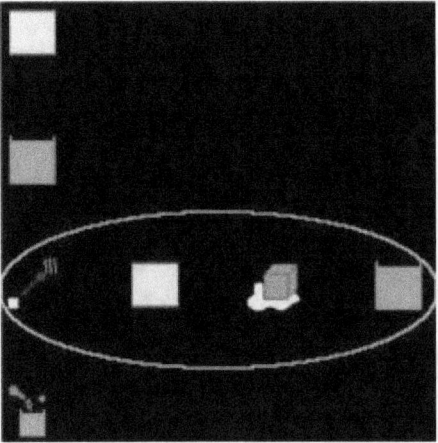

To the element Block, that represent the calorimeter of copper and gonna be called *Copper*, we have:

Name	Copper
m	105 @ g
c	390
Ti	23 @ C
Tf	Tf @ C

To the element Liquid, that represent the water into the calorimeter and gonna be called *Water*, we have:

Name	Water
m	307 @ g
c	4187

Ti	23 @ C
Tf	Tf @ C

Now, let's assume that the ice melts completely and that the resulting liquid reaches a higher temperature at 0 C. Then to the second element Block, representing the ice in the temperature range of -4 C to 0 C and gonna be called *IceA*, we have:

Name	IceA
m	95 @ g
c	2090
Ti	-4 @ C
Tf	0 @ C

Now to the element Change of state solid-liquid, representing the melting of ice and gonna be called *IceB*, we have:

Name	IceB
m	95 @ g
cf	79.7 @ cal/g
Sense	>

And to the second element Liquid, representing liquid water obtained by melting ice and gonna be called *IceC*, we have:

Name	IceC
m	95 @ g
c	4187
Ti	0 @ C
Tf	Tf @ C

Now to the element Process, which represents the total change suffered by the ice water and gonna be called *Ice*, we have:

Name	Ice
Object 1	IceA
Object 2	IceB

Object 3   IceC

Object 4   0

Object 5   0

Note that the unknown data of the final temperature is the same in all elements, as must be. Therefore, to the element Calorimeter:

Object 1   Copper

Object 2   Water

Object 3   Ice

Object 4   0

Now click in the icon Solve to get the answer:

```
Tf = -1.154 C ;
Status = success.

Verify that the calculated temperature is within the
range expected. Otherwise, some process or change never
takes place because the energy isn't enough.
```

The warning message tells us that we must verify that the final temperature is within the expected range. Obviously the obtained final temperature is not within the expected range, since we assume that the ice has completely melted and the resulting liquid water took on a final temperature above 0 C. This suggests (as tells us the warning message) that some process is not carried out. The immediate possibility is that the energy of the calorimeter and water, is not enough to melt all the ice. So let's assume that the ice reaches the 0 C, but only part of it melts. Thus removing the element Liquid, which represents the resulting water ice, and also delete its name from the element Process. Then the final temperature isn't an unknown data, because if only part of the ice melts, then the final temperature is 0 C. The new unknown data is the mass of ice that melts. Thus, with this new hypothesis the elements to the problem are as shown the image below:

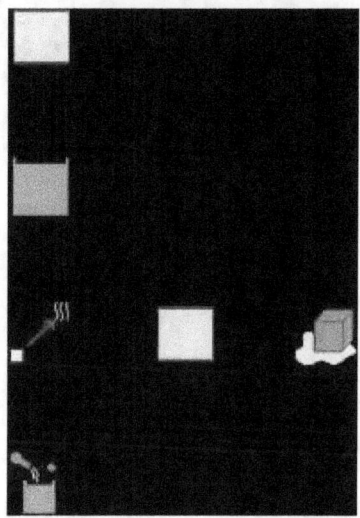

We must delete the element *IceC* in the element Process:

Name	Ice
Object 1	IceA
Object 2	IceB
Object 3	0
Object 4	0
Object 5	0

In the element Change of state solid-liquid, the mass is now the unknown data (we add the conversion factor to grams):

Name	IceB
m	m @ g
cf	79.7 @ cal/g
Sense	>

In the elements that represent the copper and its water, the final temperature is now 0 C:

Name	Copper
m	105 @ g
c	390
Ti	23 @ C
Tf	0 @ C

Name	Water
m	307 @ g
c	4187
Ti	23 @ C
Tf	0 @ C

Now click in the icon Solve to get the answer:

```
m = 89.037 g ;
Status = success.
```

Therefor, the final temperature is 0 C, with only 89 g of melting ice.

15.13 Example 13

One type of oil (specific heat 0.51 cal/g*K) is cooled in a heat exchanger. If the oil, with a flow of 3 L/min, enters to the heat exchanger at 72 C and exits with a temperature of 27 C, What is the flow of the water, used as refrigerant (specific heat 4187 J/kg*K), if this enters with a temperature of 20 C and exits with a temperature of 63 C? Give your answer on L/min.

Solution with FisicaLab

Select the Thermodynamics group and, inside this, the Calorimetry and gases module. Erase the content of the chalkboard. And add one element Heat exchanger, as show the image below:

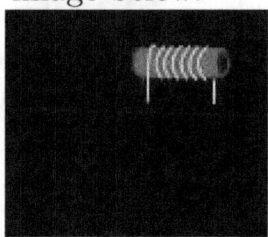

To this element:

TRi	20 @ C
TRf	63 @ C
dR/dt	Q @ L/min
cR	4187
TFi	72 @ C
TFf	27 @ C
dF/dt	3 @ L/min
cF	0.51 @ cal/g*K

Now click in the icon Solve to get the answer:

```
Q = 1.601 L/min ;
Status = success.
```

16 GNU Free Documentation License

Version 1.3, 3 November 2008

0. PREAMBLE

The purpose of this License is to make a manual, textbook, or other
functional and useful document *free* in the sense of freedom: to assure
everyone the effective freedom to copy and redistribute it, with or without
modifying it, either commercially or noncommercially. Secondarily,
this License preserves for the author and publisher a way to get credit
for their work, while not being considered responsible for modifications
made by others.

This License is a kind of "copyleft", which means that derivative works
of the document must themselves be free in the same sense. It com-
plements the GNU General Public License, which is a copyleft license
designed for free software.

We have designed this License in order to use it for manuals for free software, because free software needs free documentation: a free program
should come with manuals providing the same freedoms that the software does. But this License is not limited to software manuals; it can
be used for any textual work, regardless of subject matter or whether it
is published as a printed book. We recommend this License principally
for works whose purpose is instruction or reference.

1. APPLICABILITY AND DEFINITIONS

This License applies to any manual or other work, in any medium,
that contains a notice placed by the copyright holder saying it can be
distributed under the terms of this License. Such a notice grants a
world-wide, royalty-free license, unlimited in duration, to use that work
under the conditions stated herein. The "Document", below, refers to
any such manual or work. Any member of the public is a licensee, and
is addressed as "you". You accept the license if you copy, modify or
distribute the work in a way requiring permission under copyright law.

A "Modified Version" of the Document means any work containing the Document or a portion of it, either copied verbatim, or with modifications and/or translated into another language.

A "Secondary Section" is a named appendix or a front-matter section of the Document that deals exclusively with the relationship of the publishers or authors of the Document to the Document's overall subject (or to related matters) and contains nothing that could fall directly within that overall subject. (Thus, if the Document is in part a textbook of mathematics, a Secondary Section may not explain any mathematics.) The relationship could be a matter of historical connection with the subject or with related matters, or of legal, commercial, philosophical, ethical or political position regarding them.

The "Invariant Sections" are certain Secondary Sections whose titles are designated, as being those of Invariant Sections, in the notice that says that the Document is released under this License. If a section does not fit the above definition of Secondary then it is not allowed to be designated as Invariant. The Document may contain zero Invariant Sections. If the Document does not identify any Invariant Sections then there are none.

The "Cover Texts" are certain short passages of text that are listed, as Front-Cover Texts or Back-Cover Texts, in the notice that says that the Document is released under this License. A Front-Cover Text may be at most 5 words, and a Back-Cover Text may be at most 25 words.

A "Transparent" copy of the Document means a machine-readable copy, represented in a format whose specification is available to the general public, that is suitable for revising the document straightforwardly with generic text editors or (for images composed of pixels) generic paint programs or (for drawings) some widely available drawing editor, and that is suitable for input to text formatters or for automatic translation to a variety of formats suitable for input to text formatters. A copy made in an otherwise Transparent file format whose markup, or absence of markup, has been arranged to thwart or discourage subsequent modification by readers is not Transparent. An image format is not Transparent if used for any substantial amount of text. A copy that is not "Transparent" is called "Opaque".

Examples of suitable formats for Transparent copies include plain ASCII without markup, Texinfo input format, LaTeX input format, SGML or XML using a publicly available DTD, and standard-conforming simple HTML, PostScript or PDF designed for human modification. Examples of transparent image formats include PNG, XCF and JPG. Opaque formats include proprietary formats that can be read and edited only by proprietary word processors, SGML or XML for which the DTD and/or processing tools are not generally available, and the machine-generated HTML, PostScript or PDF produced by some word processors for output purposes only.

The "Title Page" means, for a printed book, the title page itself, plus such following pages as are needed to hold, legibly, the material this License requires to appear in the title page. For works in formats which do not have any title page as such, "Title Page" means the text near the most prominent appearance of the work's title, preceding the beginning of the body of the text.

The "publisher" means any person or entity that distributes copies of the Document to the public.

A section "Entitled XYZ" means a named subunit of the Document whose title either is precisely XYZ or contains XYZ in parentheses following text that translates XYZ in another language. (Here XYZ stands for a specific section name mentioned below, such as "Acknowledgements", "Dedications", "Endorsements", or "History".) To "Preserve the Title" of such a section when you modify the Document means that it remains a section "Entitled XYZ" according to this definition.

The Document may include Warranty Disclaimers next to the notice which states that this License applies to the Document. These Warranty Disclaimers are considered to be included by reference in this License, but only as regards disclaiming warranties: any other implication that these Warranty Disclaimers may have is void and has no effect on the meaning of this License.

2. VERBATIM COPYING

You may copy and distribute the Document in any medium, either commercially or noncommercially, provided that this License, the copyright notices, and the license notice saying this License applies to the Document are reproduced in all copies, and that you add no other conditions whatsoever to those of this License. You may not use technical measures to obstruct or control the reading or further copying of the copies you make or distribute. However, you may accept compensation in exchange for copies. If you distribute a large enough number of copies you must also follow the conditions in section 3.

You may also lend copies, under the same conditions stated above, and you may publicly display copies.

3. COPYING IN QUANTITY

If you publish printed copies (or copies in media that commonly have printed covers) of the Document, numbering more than 100, and the Document's license notice requires Cover Texts, you must enclose the copies in covers that carry, clearly and legibly, all these Cover Texts: Front-Cover Texts on the front cover, and Back-Cover Texts on the back cover. Both covers must also clearly and legibly identify you as the publisher of these copies. The front cover must present the full title with all words of the title equally prominent and visible. You may add other material on the covers in addition. Copying with changes limited to the covers, as long as they preserve the title of the Document and satisfy these conditions, can be treated as verbatim copying in other respects.

If the required texts for either cover are too voluminous to fit legibly, you should put the first ones listed (as many as fit reasonably) on the actual cover, and continue the rest onto adjacent pages.

If you publish or distribute Opaque copies of the Document numbering more than 100, you must either include a machine-readable Transparent copy along with each Opaque copy, or state in or with each Opaque copy a computer-network location from which the general network-using public has access to download using public-standard network protocols a complete Transparent copy of the Document, free of added material. If you use the latter option, you must take reasonably prudent steps, when you begin distribution of Opaque copies in quantity, to ensure that this Transparent copy will remain thus accessible at the stated location until at least one year after the last time you distribute an Opaque copy (directly or through your agents or retailers) of that edition to the public.

It is requested, but not required, that you contact the authors of the Document well before redistributing any large number of copies, to give them a chance to provide you with an updated version of the Document.

4. MODIFICATIONS

You may copy and distribute a Modified Version of the Document under the conditions of sections 2 and 3 above, provided that you release the Modified Version under precisely this License, with the Modified Version filling the role of the Document, thus licensing distribution and modification of the Modified Version to whoever possesses a copy of it. In addition, you must do these things in the Modified Version:

A. Use in the Title Page (and on the covers, if any) a title distinct from that of the Document, and from those of previous versions (which should, if there were any, be listed in the History section of the Document). You may use the same title as a previous version if the original publisher of that version gives permission.

B. List on the Title Page, as authors, one or more persons or entities responsible for authorship of the modifications in the Modified Version, together with at least five of the principal authors of the Document (all of its principal authors, if it has fewer than five), unless they release you from this requirement.

C. State on the Title page the name of the publisher of the Modified Version, as the publisher.

D. Preserve all the copyright notices of the Document.

E. Add an appropriate copyright notice for your modifications adjacent to the other copyright notices.

F. Include, immediately after the copyright notices, a license notice giving the public permission to use the Modified Version under the terms of this License, in the form shown in the Addendum below.

G. Preserve in that license notice the full lists of Invariant Sections and required Cover Texts given in the Document's license notice.

H. Include an unaltered copy of this License.

I. Preserve the section Entitled "History", Preserve its Title, and add to it an item stating at least the title, year, new authors, and publisher of the Modified Version as given on the Title Page. If there is no section Entitled "History" in the Document, create one stating the title, year, authors, and publisher of the Document as given on its Title Page, then add an item describing the Modified Version as stated in the previous sentence.

J. Preserve the network location, if any, given in the Document for public access to a Transparent copy of the Document, and likewise the network locations given in the Document for previous versions it was based on. These may be placed in the "History" section. You may omit a network location for a work that was published at least four years before the Document itself, or if the original publisher of the version it refers to gives permission.

K. For any section Entitled "Acknowledgements" or "Dedications", Preserve the Title of the section, and preserve in the section all the substance and tone of each of the contributor acknowledgements and/or dedications given therein.

L. Preserve all the Invariant Sections of the Document, unaltered in their text and in their titles. Section numbers or the equivalent are not considered part of the section titles.

M. Delete any section Entitled "Endorsements". Such a section may not be included in the Modified Version.

N. Do not retitle any existing section to be Entitled "Endorsements" or to conflict in title with any Invariant Section.

O. Preserve any Warranty Disclaimers.

If the Modified Version includes new front-matter sections or appendices that qualify as Secondary Sections and contain no material copied from the Document, you may at your option designate some or all of these sections as invariant. To do this, add their titles to the list of Invariant Sections in the Modified Version's license notice. These titles must be distinct from any other section titles.

You may add a section Entitled "Endorsements", provided it contains nothing but endorsements of your Modified Version by various parties—for example, statements of peer review or that the text has been approved by an organization as the authoritative definition of a standard.

You may add a passage of up to five words as a Front-Cover Text, and a passage of up to 25 words as a Back-Cover Text, to the end of the list of

Cover Texts in the Modified Version. Only one passage of Front-Cover Text and one of Back-Cover Text may be added by (or through arrangements made by) any one entity. If the Document already includes a cover text for the same cover, previously added by you or by arrangement made by the same entity you are acting on behalf of, you may not add another; but you may replace the old one, on explicit permission from the previous publisher that added the old one.

The author(s) and publisher(s) of the Document do not by this License give permission to use their names for publicity for or to assert or imply endorsement of any Modified Version.

5. COMBINING DOCUMENTS

You may combine the Document with other documents released under this License, under the terms defined in section 4 above for modified versions, provided that you include in the combination all of the Invariant Sections of all of the original documents, unmodified, and list them all as Invariant Sections of your combined work in its license notice, and that you preserve all their Warranty Disclaimers.

The combined work need only contain one copy of this License, and multiple identical Invariant Sections may be replaced with a single copy. If there are multiple Invariant Sections with the same name but different contents, make the title of each such section unique by adding at the end of it, in parentheses, the name of the original author or publisher of that section if known, or else a unique number. Make the same adjustment to the section titles in the list of Invariant Sections in the license notice of the combined work.

In the combination, you must combine any sections Entitled "History" in the various original documents, forming one section Entitled "History"; likewise combine any sections Entitled "Acknowledgements", and any sections Entitled "Dedications". You must delete all sections Entitled "Endorsements."

6. COLLECTIONS OF DOCUMENTS

You may make a collection consisting of the Document and other documents released under this License, and replace the individual copies of this License in the various documents with a single copy that is included in the collection, provided that you follow the rules of this License for verbatim copying of each of the documents in all other respects.

You may extract a single document from such a collection, and distribute it individually under this License, provided you insert a copy of this License into the extracted document, and follow this License in all other respects regarding verbatim copying of that document.

7. AGGREGATION WITH INDEPENDENT WORKS

A compilation of the Document or its derivatives with other separate and independent documents or works, in or on a volume of a storage or distribution medium, is called an "aggregate" if the copyright resulting from the compilation is not used to limit the legal rights of the compilation's users beyond what the individual works permit. When the Document is included in an aggregate, this License does not apply to the other works in the aggregate which are not themselves derivative works of the Document.

If the Cover Text requirement of section 3 is applicable to these copies of the Document, then if the Document is less than one half of the entire aggregate, the Document's Cover Texts may be placed on covers that bracket the Document within the aggregate, or the electronic equivalent of covers if the Document is in electronic form. Otherwise they must appear on printed covers that bracket the whole aggregate.

8. TRANSLATION

Translation is considered a kind of modification, so you may distribute translations of the Document under the terms of section 4. Replacing Invariant Sections with translations requires special permission from their copyright holders, but you may include translations of some or all Invariant Sections in addition to the original versions of these Invariant Sections. You may include a translation of this License, and all the license notices in the Document, and any Warranty Disclaimers, provided that you also include the original English version of this License and the original versions of those notices and disclaimers. In case of a disagreement between the translation and the original version of this License or a notice or disclaimer, the original version will prevail.

If a section in the Document is Entitled "Acknowledgements", "Dedications", or "History", the requirement (section 4) to Preserve its Title (section 1) will typically require changing the actual title.

9. TERMINATION

You may not copy, modify, sublicense, or distribute the Document except as expressly provided under this License. Any attempt otherwise to copy, modify, sublicense, or distribute it is void, and will automatically terminate your rights under this License.

However, if you cease all violation of this License, then your license from a particular copyright holder is reinstated (a) provisionally, unless and until the copyright holder explicitly and finally terminates your license, and (b) permanently, if the copyright holder fails to notify you of the violation by some reasonable means prior to 60 days after the cessation.

Moreover, your license from a particular copyright holder is reinstated permanently if the copyright holder notifies you of the violation by some reasonable means, this is the first time you have received notice of violation of this License (for any work) from that copyright holder, and you cure the violation prior to 30 days after your receipt of the notice.

Termination of your rights under this section does not terminate the licenses of parties who have received copies or rights from you under this License. If your rights have been terminated and not permanently reinstated, receipt of a copy of some or all of the same material does not give you any rights to use it.

10. FUTURE REVISIONS OF THIS LICENSE

The Free Software Foundation may publish new, revised versions of the GNU Free Documentation License from time to time. Such new versions will be similar in spirit to the present version, but may differ in detail to address new problems or concerns. See `http://www.gnu.org/copyleft/`.

Each version of the License is given a distinguishing version number. If the Document specifies that a particular numbered version of this License "or any later version" applies to it, you have the option of following the terms and conditions either of that specified version or of any later version that has been published (not as a draft) by the Free Software Foundation. If the Document does not specify a version number of this License, you may choose any version ever published (not as a draft) by the Free Software Foundation. If the Document specifies that a proxy can decide which future versions of this License can be used, that proxy's public statement of acceptance of a version permanently authorizes you to choose that version for the Document.

11. RELICENSING

"Massive Multiauthor Collaboration Site" (or "MMC Site") means any World Wide Web server that publishes copyrightable works and also provides prominent facilities for anybody to edit those works. A public wiki that anybody can edit is an example of such a server. A "Massive Multiauthor Collaboration" (or "MMC") contained in the site means any set of copyrightable works thus published on the MMC site.

"CC-BY-SA" means the Creative Commons Attribution-Share Alike 3.0 license published by Creative Commons Corporation, a not-for-profit corporation with a principal place of business in San Francisco, California, as well as future copyleft versions of that license published by that same organization.

"Incorporate" means to publish or republish a Document, in whole or in part, as part of another Document.

An MMC is "eligible for relicensing" if it is licensed under this License, and if all works that were first published under this License somewhere other than this MMC, and subsequently incorporated in whole or in part into the MMC, (1) had no cover texts or invariant sections, and (2) were thus incorporated prior to November 1, 2008.

The operator of an MMC Site may republish an MMC contained in the site under CC-BY-SA on the same site at any time before August 1, 2009, provided the MMC is eligible for relicensing.

ADDENDUM: How to use this License for your documents

To use this License in a document you have written, include a copy of the License in the document and put the following copyright and license notices just after the title page:

```
Copyright (C)  year  your name.
Permission is granted to copy, distribute and/or modify this document
under the terms of the GNU Free Documentation License, Version 1.3
or any later version published by the Free Software Foundation;
with no Invariant Sections, no Front-Cover Texts, and no Back-Cover
Texts.  A copy of the license is included in the section entitled ''GNU
Free Documentation License''.
```

If you have Invariant Sections, Front-Cover Texts and Back-Cover Texts, replace the "with...Texts." line with this:

```
with the Invariant Sections being list their titles, with
the Front-Cover Texts being list, and with the Back-Cover Texts
being list.
```

If you have Invariant Sections without Cover Texts, or some other combination of the three, merge those two alternatives to suit the situation.

If your document contains nontrivial examples of program code, we recommend releasing these examples in parallel under your choice of free software license, such as the GNU General Public License, to permit their use in free software.

Indice

A

Absolute velocity (cd) 203
Angles (cd) 199
Angles (rs) 87
Angles (s) 61
Angular acceleration (cd) 191
Angular acceleration (ck) 40
Angular momentum (cd) 192
Angular velocity (cd) 190
Angular velocity (ck) 40
Applied heat (cal) 260
Applied heat flow (cal) 260
Arc length (ck) 44

B

Beam (rs) 85
Beams of 2 forces (rs) 92
Beams of truss (rs) 94
Block (cal) 262
Block (s) 58
Block above an inclined plane to the left (s) 59
Block above an inclined plane to the right (s) 59
Block with horizontal movement (d) .. 141
Block with movement along an inclined plane to the left (d) 142
Block with movement along an inclined plane to the right (d) 143
Block with vertical movement (d) 140

C

Calorimeter (cal) 268
Calorimetry 258
Cannon (k) 16
Center of rotation (cd) 197
Center of rotation (ck) 43
Centripetal acceleration (cd) 190
Change of state liquid-gas (cal) 266
Change of state solid-liquid (cal) 265
Collision (d) 148
Coordinate (ck) 44
Couple (rs) 91

D

Dilatación superficial (cal) 264
Distance (ck) 43
Distance (k) 18
Distance XY (k) 18

E

Element data 4
Elementos de viga (rs) 88
Elements of solid (rs) 89
Energy (cd) 191
Energy (d) 149
Example 1 (cal) 272
Example 1 (cd) 210
Example 1 (ck) 46
Example 1 (d) 151
Example 1 (k) 20
Example 1 (rs) 97
Example 1 (s) 64
Example 10 (cal) 284
Example 10 (cd) 246
Example 10 (d) 174
Example 10 (rs) 120
Example 11 (cal) 286
Example 11 (cd) 251
Example 11 (d) 176
Example 11 (rs) 125
Example 12 (cal) 288
Example 12 (cd) 253
Example 12 (d) 180
Example 12 (rs) 127
Example 13 (cal) 293
Example 13 (cd) 256
Example 13 (rs) 131
Example 2 (cal) 273
Example 2 (cd) 213
Example 2 (ck) 48
Example 2 (d) 153
Example 2 (k) 22
Example 2 (rs) 100
Example 2 (s) 66
Example 3 (cal) 275
Example 3 (cd) 216
Example 3 (ck) 50
Example 3 (d) 155
Example 3 (k) 24
Example 3 (rs) 102

Example 3 (s) 68
Example 4 (cal) 276
Example 4 (cd) 219
Example 4 (ck) 52
Example 4 (d) 158
Example 4 (k) 26
Example 4 (rs) 106
Example 4 (s) 70
Example 5 (cal) 277
Example 5 (cd) 224
Example 5 (ck) 54
Example 5 (d) 162
Example 5 (k) 29
Example 5 (rs) 108
Example 5 (s) 72
Example 6 (cal) 278
Example 6 (cd) 228
Example 6 (d) 165
Example 6 (k) 31
Example 6 (rs) 111
Example 6 (s) 74
Example 7 (cal) 280
Example 7 (cd) 233
Example 7 (d) 167
Example 7 (k) 33
Example 7 (rs) 113
Example 7 (s) 76
Example 8 (cal) 282
Example 8 (cd) 237
Example 8 (d) 169
Example 8 (rs) 115
Example 8 (s) 78
Example 9 (cal) 283
Example 9 (cd) 242
Example 9 (d) 172
Example 9 (rs) 117
Example 9 (s) 81
Examples calorimetry 272
Examples circular dynamics of particles
    210
Examples circular kinematics of particles
    46
Examples dynamics of particles 151
Examples kinematics of particles 20
Examples statics 64
Examples statics rigid bodies 97

F

Final System (cd) 196
Forces (cd) 198
Forces (d) 145

Forces (rs) 90
Forces (s) 62
Frequency (ck) 41
Fricciones (rs) 90
Frictions (cd) 199
Frictions (d) 145
Frictions (s) 62
Frictions between blocks (contacts) (d)
    146

G

Gas (cal) 263
Gas at constant pressure (cal) 269
Gas at constant temperature (cal) 269
Gas at constant volume (cal) 270
Gas PV/T (cal) 270
GNU Free Documentation License 294

H

Handling the elements 3
Heat exchanger (cal) 271
Heat extracted (cal) 261
How it works 7

I

Inertia (cd) 202
Initial System (cd) 195
Introduction 1

J

Joint (rs) 93

L

Laboratory clock (cal) 260
Linear expansion (cal) 263
Linear momentum (cd) 193
Liquid (cal) 262

M

Maximum acceleration (cd) 201
Messages 7
Mobile (d) 138
Mobile (k) 13
Mobile in X/Y (d) 139
Mobile in X/Y (k) 14

Mobile in X/Y with constant velocity (k)
.... 15
Mobile radial (k) 17
Mobile reference system (k) 11
Mobile reference system in X/Y (k) ... 12
Mobile with circular motion (ck) 38
Mobile with circular movement (cd) .. 187
Mobile with linear movement (cd) 186
Mobile with perpendicular circular
       movement (cd) 189
Mobile with polar circular motion (ck)
.... 39
Mobile with polar circular movement (cd)
.... 188
Module circular dynamics of particles
.... 182
Module circular kinematics of particles
.... 35
Module dynamics of particles........ 135
Module kinematics of particles........ 8
Module rigid statics 83
Module statics of particles 57
Moment of a force or couple of forces (cd)
.... 200
Momentum (d) 150

N
Number of laps (ck) 42

O
Object in rest (cd) 185

P
Period (ck)........................... 42
Point (k)............................. 19
Point (rs) 87
Points (rs) 84
Power (cd)........................... 194
Power (d)............................ 150
Process (cal) 267
Pulley (d)............................ 144

Pulley (s) 60

R
Refrigeration (cal) 261
Relation between accelerations (d).... 146
Relative motion (d) 147
Relative velocity (ck)................. 45
Relative velocity (k).................. 19
Resultant (rs) 95
Resultant (s).......................... 63
Resultante con fuerza horizontal (rs) .. 95
Resultante con fuerza vertical (rs) 96

S
Sine of angle (cd) 204
Solid (rs) 86
Springs (cd) 198
Springs (d) 144
Springs (s)............................ 60
Static point (s)....................... 61
Stationary reference system (cd)...... 185
Stationary reference system (ck)....... 37
Stationary reference system (d)....... 137
Stationary reference system (k)........ 10
Stationary reference system (rs) 84
Stationary reference system (s) 58
Supported combinations of elements
       Energy, Angular Momentum, Linear
       Momentum and Power (cd) 205

T
Total acceleration (ck) 41
Total acceleration (Triangle of
       accelerations) (cd) 201
Truss (rs) 93

V
Vertical/Horizontal resultant 63
Volumetric expansion (cal) 264

www.ingramcontent.com/pod-product-compliance
Lightning Source LLC
Chambersburg PA
CBHW062348220526
45472CB00008B/1742